Indigenous Sustainable Wisdom

This book is part of the Peter Lang Education list.
Every volume is peer reviewed and meets
the highest quality standards for content and production.

PETER LANG
New York • Bern • Berlin
Brussels • Vienna • Oxford • Warsaw

Indigenous Sustainable Wisdom

First-Nation Know-How for Global Flourishing

Edited by
Darcia Narvaez,
Four Arrows (Don Trent Jacobs),
Eugene Halton, Brian S Collier,
and Georges Enderle

PETER LANG
New York • Bern • Berlin
Brussels • Vienna • Oxford • Warsaw

Library of Congress Cataloging-in-Publication Data
Names: Narvaez, Darcia, editor.
Title: Indigenous sustainable wisdom: first-nation know-how for
global flourishing / edited by Darcia Narvaez, Four Arrows (Don Trent Jacobs),
Eugene Halton, Brian S Collier, and Georges Enderle.
Description: New York: Peter Lang, 2019.
Includes bibliographical references and index.
Identifiers: LCCN 2018052216 | ISBN 978-1-4331-6365-4 (hardback: alk. paper)
ISBN 978-1-4331-6364-7 (paperback: alk. paper)
ISBN 978-1-4331-6008-0 (ebook pdf)
ISBN 978-1-4331-6009-7 (epub) | ISBN 978-1-4331-6010-3 (mobi)
Subjects: LCSH: Indigenous peoples. | Sustainable development—
Social aspects. | Globalization—Social aspects. | Culture and globalization.
Classification: LCC GN380.I538 2019 | DDC 305.8—dc23
LC record available at https://lccn.loc.gov/2018052216
DOI 10.3726/b14900

Bibliographic information published by **Die Deutsche Nationalbibliothek**.
Die Deutsche Nationalbibliothek lists this publication in the "Deutsche
Nationalbibliografie"; detailed bibliographic data are available
on the Internet at http://dnb.d-nb.de/.

© 2019 Peter Lang Publishing, Inc., New York
29 Broadway, 18th floor, New York, NY 10006
www.peterlang.com

We dedicate the energy and contributions gathered here to the wellbeing of the next seven generations, and beyond.

Contents

Figures

Acknowledgments

We thank the courageous Indigenous Peoples, often unrecognized, who have fought against all odds to maintain humanity's original ways of knowing. We especially thank the generous support of The Pokagon Band of the Potawatomi (*Pokégnek Bodéwadmik)* upon whose traditional homeland, for thousands of years, the University of Notre Dame is placed. This book (and the conference from which it emerged) was also made possible in part by support from the Institute for Scholarship in the Liberal Arts, College of Arts and Letters, University of Notre Dame and from many entities across the university. We thank the following units at the University of Notre Dame:

Center for Arts and Cultures
Center for Social Concerns
College of Science, Nieuwland Lecture Series
Department of American Studies
Department of Anthropology
Department of Art, Art History & Design
Department of English
Department of History
Department of Psychology
Department of Sociology

Department of Theology
Eck Institute for Global Health
Environmental Change Initiative
Gender Studies Program
The Graduate School
Institute for Advanced Study
Institute for Educational Initiatives
John J. Reilly Center for Science, Technology & Values
Kroc Institute for International Peace Studies
The Law School The Minor in Sustainability
Multicultural Student Programs and Services
Native American Alumni
Native American Initiatives
Native American Student Association of Notre Dame
Office of Research
Poverty Studies
Shaw Center for Children and Families
The Snite Museum

CHAPTER ONE

People and Planet in Need of Sustainable Wisdom

DARCIA NARVAEZ, FOUR ARROWS, EUGENE HALTON,
BRIAN S COLLIER AND GEORGES ENDERLE

It is always a shock to learn of the teeming wildlife that greeted the first Western explorers and invaders of the Americas a few centuries ago. Columbus and others who followed him reported on the sweet smells of blossoms and trees from hundreds of miles off the coast, even off the Jersey shore.[1] It was a land of luxuriant vegetation filled with huge herds of animals and flocks of birds that could darken the sky for hours if not days during migrations. In the interior, romps of river otters and knots of giant catfish impeded the progress of canoes. The land gave forth a profusion of life that seems impossible to imagine today. Yet how we might envision and bring about such a return to biodiversity, as well as cultural diversity, is one of the questions this book attempts to address.

Today we may be concerned about the destruction of the Amazon rainforest and other remaining pristine forests, but will our actions on these concerns be as inadequate as in the past? For example, in the 19th century, the cutting and burning of the old growth forests across the United States proceeded at a rate of up to 25 million acres a year, with regular celebrated intentional massacres of wildlife—from flocks of passenger pigeons and Carolina parakeets (which went extinct as a result) to millions of bison. In 1818, citizens in Hinckley Township Ohio came together in the nearby forest to kill hundreds of "pests": "three hundred deer, twenty-one bears, seventeen wolves, and uncounted numbers of foxes and other small game."[2] In a brief amount of time, the invaders transformed the vast paradise into a deforested, parched, and eroded landscape. The ravaging included old growth cultures—at least 600 different societies existed across what became the United States.[3]

California offers a relatively recent example of such extermination practices that included not only plants and animals, but humans. Prior to European contact in 1543 and stretching back earlier, through at least 12,000 years of mostly peaceful cooperation, California was home to approximately 70 native tribes, speaking about 100 languages from 5 different language families, as well as vast herds of antelope, deer, and elk, rivers teeming with fish, coastlines banked by thousands of sea otters and other sea life, and massive old growth forests.[4] Thousands of native peoples were lost initially through systematic enslavement and torture during Spanish occupation from 1769 until 1821 when Mexico won its independence from Spain. The U.S. military started to invade California in 1846, before the ultimate transfer to the United States in 1848, when a systematic hunting and slaughtering of natives by vigilantes and soldiers began. Many genocidal actions were directed by governors and congressmen, with the feverish lure of the gold rush as a backdrop. By 1873, only 30,000 native persons were known to remain (though many others likely persisted, by passing as non-native). Much of California's ecological wealth was also decimated during the 19th century.

The mindset and practices that exterminated species and cultures in the past are still with us today. In fact, in recent decades, they have spread to virtually every corner of the earth.[5] Moreover, the prognosis is not good as global warming and decimation of ecosystems accelerates in the 21st century. This book offers new attention to alternative mindsets and practices that are based in sustainable cultural wisdom.

The Setting

While the 21st century has seen the growth of a staggering amount of financial wealth around the world, inequality has burgeoned. Attacks on bio- and cultural-diversity continue unabated, despite increased knowledge of the interrelations among and within ecological systems. Increasing technological capacities have only worsened the situation. Signs of ill-being among ecosystems, biodiversity, and plant and animal life have surged.[6] The dominant culture of the earth, whose roots grow out of earlier Middle Eastern empires and patriarchal religions, exploded out of Europe with globalizing imperialism, capitalism, and technological power. Now fully globalized, it is putting all life on the planet at risk.[7] The dominant worldview, largely driven by the fever of financial wealth within economically-advanced nations in the northern hemisphere, has emphasized financial wealth at the expense of social and ecological wealth and health.[8] The Dakota pipeline project—which gave rise to the Standing Rock protest led by the Lakota, Dakota and Nakota People—is one of many examples of such hegemonic priorities. In 2016, Energy Transfer Partners, in concert with the federal government, ignored long-standing treaty obligations and well-founded concerns regarding the potential for oil leaking into the drinking water of over ten million people in order to lay oil pipe across the sacred Missouri River and tribal lands.

These same tribes were all too familiar with such violations. In the 1950s, similar policies were responsible for placing dams on the Missouri River that selectively inundated their riparian homeland. Entire tribal communities were relocated to barren plains above the river valley.[9] This kind of displacement of people and alteration of environments in which people dwell is the story of civilization since its beginnings, undergirded by a narrative, until recently, of continued progress, but one that masked its real underlying costs. Blind and destructive "progress" is a cause for great biocultural destruction, as noted in the 2015 papal encyclical letter, *Laudato Si*,

> in various parts of the world, pressure is being put on [Indigenous communities] to abandon their homelands to make room for agricultural or mining projects which are undertaken without regard for the degradation of nature and culture. Authentic development includes efforts to bring about an integral improvement in the quality of human life, and this entails considering the setting in which people live their lives.[10]

The rise of the modern world, rooted in domination of Nature and technological progress, has increasingly revealed its shortcomings and unsustainability through human-caused cultural, ecological, and biodiversity destruction.[11]

As mentioned earlier, even in advanced nations like the United States, the situation is worsening despite overall vast financial wealth.[12] This caps a long-term trend showing an inverse relationship between financial wealth and wellbeing. Physical anthropologist Clark Spencer Larsen stated: "Although agriculture provided the economic basis for the rise of states and development of civilizations, the change in diet and acquisition of food resulted in a decline in quality of life for most human populations in the last 10,000 years."[13] Western science and scholarship may lament the resulting ill health (mental and physical) and environmental destruction from their civilization's development, but all too often these are considered to be a part of "the human condition," a "necessary collateral damage," or "price of progress" which cannot be escaped. This argument suggests that "there is no other way" because, before hegemonic civilization, humanity is believed to have lived in conditions of "brutality and savagery"—in spite of the fact that all first contact accounts between Europeans and natives of the "New World" suggest otherwise—and no one wants to return to that.

However one measures inclusive wellbeing, the human species as we know it could be headed for extinction as the dominant culture leads more and more people to behave suicidally in lifestyle decisions and actions. The warning signs of a planet in distress brought out the Cogi People of Colombia, who lived in seclusion for centuries. They contacted the BBC in 1990 to convey a message to the "younger brother" (those of Western civilization) to stop destroying the earth. They could discern from their mountaintop home that the natural world was in peril. Their ethics come from the land itself: "The Great Mother taught us right and wrong. Now they are digging up the Mother's heart, and her eyes and ears. Stop digging and digging. Do not cut down trees—they hurt, like cutting off your own leg."[14] "If crops aren't properly blessed, they dry up ... that's how it is."[15] "If we act well, the earth will survive. If we do not, it will not."[16] More recently, awareness of imminent peril has found form in the emerging scientific concept of the "Anthropocene," of significant human-influenced effects on global ecosystems; a potential sixth major period of extinction is already underway.[17] Derrick Jensen has suggested that the current era be called "Sociopocene" to reflect the sociopathy that humanity is displaying in face of its destructiveness.[18] Others have suggested "Capitalocene" because capitalism is the primary culprit behind the rapid race toward planetary destruction.[19]

The Illness

What accounts for the differences between dominant global modern culture and the cultures of successful, sustainable Indigenous communities that existed for tens of thousands of years and that are still represented today in First Nations who struggle against the odds around the world? First, there are significant contrasts between the traditional Indigenous Worldview that led humanity for most of its history and the one that has become dominant over the past eight or nine thousand years.[20]

The ancient worldview considers the cosmos to be "unified, sacred and moral."[21] Communities who live with this worldview are connected to the lifeways of a particular landscape, they promote the flourishing of the local biocommunity, and view the human species as just one member of that community: one member among many members.[22] Animals are often considered as persons with lives similar to humans. In this view, all human actions directly affect the wellbeing of the biocommunity and so must be thoughtful and respectful of the other-than-human.

The other, more recent, worldview considers the cosmos "fragmented, disenchanted, amoral."[23] Modern civilization has become infused with this worldview, which focuses on preventing bad outcomes (for humans) through promoting accumulation (hoarding, then consumerism), using coercion to enforce norms of hierarchy and obedience to the system; at the same time, its efforts intentionally work against nature and focus on human aggrandizement.[24] The goal became that of ensuring, no matter the cost, that the dominant human group would survive. Often this entails exterminating anything in the way of human dominance and control—presumed necessary to avoid human suffering and death.

Second, embedded in this recent worldview, is a sense of separation from and superiority to Nature, a view that reaches all the way back to the origins of mono-agricultural civilizations. This harmful cocktail of separation and superiority has intensified in recent centuries. In fact, Western scholarship was founded on assumptions about human separation from and superiority to Nature, easily casting off the wellbeing and even continuing existence of other life forms as unimportant or collateral damage to humanity's goals. "Personhood" was removed from all but humans, and nature became a commodity for human interests.[25] The pursuit of wealth as an end in itself has resulted in unsustainable ways of living that contrast starkly with the ways of non-civilized cultures that thrived for thousands if not tens of thousands of years before being displaced.

Third, from their beginnings in the 16th century, Western science, technology, and economics have grown as outlooks based on extreme abstraction.[26] These fields advocate systematic detachment from the earth, thereby breaking the bonds of relational responsibility to other-than-humans who are studied as objects. On the other hand, detachment from relational commitment to the wellbeing of the natural world has led to sophisticated technologies, some helpful and some destructive. Detached science and technology, in part because of their philosophy of separation and control, have led to great physical comforts—but only for a minority of humans, and often with catastrophic hidden costs.

Fourth, and perhaps most tragically, Western expansionism and global control of most areas of the earth have impaired capacities to perceive alternatives to the current pathway of increased control and extermination of nature. As Thomas Berry has pointed out, humans continue to be enchanted with technology despite the fact that the "wonderworld" promised has led instead to a "waste-world."[27] Cultures that do not conform have been forced into submission to the dominant system. Landscapes have been degraded and species lost for the sake of material accumulation and "progress." Globalized culture often seems unable to perceive an alternative to the current pathway of increased control, domination of nature, and the destruction of cultures that do not conform to the dominant format.

When considered from the perspective of the history of human cultures, predominant Western beliefs of individualism, human superiority, and separation from nature, along with their resulting practices, appear odd, rare, and even aberrant.[28] Most societies in the history of the world would consider individual "self-interest"—assumed to be normal human nature in most of the West—to be profoundly destructive and even a sign of insanity.

Indigenous World View

Non-industrialized, First-Nation, Indigenous societies around the world support the ancient worldview, an "Indigenous" worldview that can provide a portal to global flourishing.[29] An emphasis on "global flourishing" (in contrast to "economic globalization") provides an alternative to the fatalistic pathway of the dominant worldview. The goal is to promote a flourishing life through following nature's gift economy (constant sharing, reciprocal giving-and-taking among entities), being receptive to the natural flow of events, working with nature respectfully to promote diversity, including valuing human life as inseparable from the lives of other members of the bio-community. First-Nation societies "feel themselves to be guests not masters."[30] They display a whole different awareness of humanity's place: walking *with* the earth, not simply on it, and walking within its relational grasp.[31] Accordingly, the view includes a sense that spirit pervades all things, that All are related and indeed, that humans are younger siblings who have much to learn from creatures who have resided on this planet far longer. This worldview is accompanied by a sense of place on the earth, by a feeling of being at home, and by practices of fitting in with the local landscape and biocommunity.

Many First-Nations peoples of the earth have lived well and kindly with the earth for generations. Many belong to cultures around the world that lived sustainably and relatively peacefully for tens of thousands of years.[32] Their companionship orientation—from raising children, to living with humans and non-humans—fosters enduring wisdom, morality and flourishing.[33] These societies are continuous with 99% of the history of the genus *homo* and provide evidence for humanity's sustainable relationship with the rest of Nature.[34] Thus, we may regard the Indigenous worldview and lifestyle as a baseline for humanity living well on the earth and as an alternative pathway for human futures.[35]

Broken Pathways

This book addresses a number of reasons that we have departed from humanity's original—Indigenous—worldview. Paul Shepard notes a fundamental reason that becomes apparent throughout the chapters in this book. Shepard describes the broken pathway to adulthood that plagues financially wealthy nations by throwing a child off the pathway to maturity.[36] More recently, the epigenetic and developmental neurobiological outcomes of this altered child-to-adult pathway have been delineated by scholars who have noted the power of early life experience on neurobiological and social capacities, as well as on worldview.[37] In many Western societies, there has been a divorce between adult behavior and the development of wellbeing in children—a blindness about the immaturity of humans at birth and ignorance of humans as dynamic systems who require a particular species-typical nest to foster human nature and potential, one which includes lengthy, intensive support to reach maturity.[38] Most recently, Darcia Narvaez has suggested that the missing species-typical nest plays a large role in undermining sense and sensibility in adulthood, including cultural assumptions about the natural world.[39] When the species-typical nest is missing, individuals misdevelop in the ways apparent and widespread in advanced nations. Often they are restless and unattached to the local other-than-human world and focused on self-protective self-aggrandizement (financial self-interest) at the expense of future generations.[40] Detachment from close connection (parents to children, family to family member; human to other-than-humans) is now built into U.S. institutions and systems.[41] The received view normalizes the view that human nature is innately competitive and selfish and collateral destruction cannot be helped.

The misunderstandings of human nature and history is especially embedded into the psyche of America, beginning perhaps with the view that the Americas were a wilderness brought under proper control by European settlers. Through a selective re-telling of history, it came to be believed that those who lived in the Americas before European settlement either were savage (and evil), spiritually "primitive," or undeserving of the land because they did not manage it in the proper, European, agricultural way (visible manipulation and coercion of other-than-humans for human ends).[42] They were considered part of Nature, and nature was to be exploited. At the same time, U.S. history books tend to discuss First Nations as relics or extinct ("firsting and lasting"), "wiped out" by the progressive wave from European expansion.[43] However, in fact North America was not a wilderness but inhabited, nourished and enhanced by small-scale ingenious innovations with a partially cultivated landscape.[44] Many American First-Nation peoples still exist (and flourish) today, continuing longstanding relations in the environments they have inhabited for thousands of years. In fact, much of Western scholarship has ignored the vast numbers of societies and perspectives that fall outside of dominant Western notions of human life, those that lived sustainably for thousands if not tens of thousands of years, even though they make a better baseline for human nature and functioning than industrialized peoples.[45]

Moving toward Sustainable Wisdom

In order to save the human species and many others with it, the focus on domination of Nature and technological progress needs to change. As noted, globalized finance and power structures have been built on shutting out responsible relations with the natural world. The frame of mind that dominates influential institutions (business, industry, military, education, politics, religion) continues to objectify Nature and deplete biocultural diversity around the world. However, science is coming to agree with the Indigenous perspective that nature is cooperative through and through—e.g., finding cooperation across species in forests, ecosystems generally, and even the microbiome within every human body—with competition a real yet minor component.[46] Science and other disciplines are starting to better appreciate how the sustainable outlooks of "First ways" practices are rooted in an ecological mindset that fosters landscape flourishing and cultural diversity.[47] Still, there remains a disconnect when it comes to the possibilities of sustainable wisdom. The modern world has much to learn from Indigenous peoples. Their wise ways and relational lifestyles can provide new frameworks for establishing sustainability more generally.[48] Native American[49] peoples are reinvigorating traditional practices that provide insight into how sustainable, respectful human cultures have profound importance for the contemporary world.[50]

Western views are that nature can only thrive by humans leaving it completely alone. This is a misconception. The flourishing landscapes in the Americas were highly "managed" by native groups through controlled burning and other methods. For example, the Coast Miwok in Marin County, California, harvested from eleven major clam beds (basically semi-domesticating them by moving them around and decreasing densities). Interestingly, following the view that leaving natural systems "alone" is the best way to preserve it, the state's Department of Fish and Game stopped the harvesting in the 1920s in order to "protect the resource." The clam population crashed[51]—a demonstration of the limitations of dominant science views which are typically closed to the deep experience-based wisdom of Indigenous science. Native American traditions approach the natural world with reciprocal, respectful relations. This includes following the partnership principles of "the honorable harvest" such as asking permission to take a life, taking only what is needed, and taking no more than half.[52]

The book brings together an interdisciplinary set of scholars ready to disseminate First-Nation wisdom in order that it might be integrated with mainstream contemporary understandings in order to move toward a flourishing earth. We take Paul Shepard's words as a guide:

> A journey to our primal world may bring answers to our ecological dilemmas … White European/Americans cannot become Hopis or Kalahari Bushmen or Magdalenian bison hunters, but elements in those cultures can be recovered or re-created because they fit the heritage and predilection of the human genome everywhere, a genome tracing back to a common ancestor that Anglos share with Hopis and Bushmen and

all the rest of Homo sapiens. The social, ecological, and ideological characteristics natural to our humanity are to be found in the lives of foragers.

Must we build a new twenty-first-century society corresponding to a hunting/ gathering culture? Of course not; humans do not consciously make cultures. What we can do is single out those many things, large and small, that characterized the social and cultural life of our ancestors—the terms under which our genome itself was shaped—and incorporate them as best we can by creating a modern life around them. We take our cues from primal cultures, the best wisdom of the deep desires of the genome. We humans are instinctive culture makers; given the pieces, the culture will reshape itself.[53]

As we reach the limits of virtually every ecological system on the earth, a *feeling* realization is emerging of the dangers and limits of Western thinking, practices and institutions.[54] The modern world as promulgated by Western minds and egos has moved hegemonically—by means of institutional religion, science, business and politics—against peoples and cultures around the world as an extension of its self-perceived mandate to dominate and control nature. Instead of solving challenges to human and earth wellbeing, technological advance and its accompanying mindsets have undermined them. Egos unchecked in business, science, and politics too often have led to unquestioning certainty of dominant beliefs (e.g., more is better) and self-focused actions (e.g., it's my right)—as if these orientations are not destructive of a biodiverse planet and moral responsibility for life on the planet is someone else's concern. This book will address how Indigenous approaches to communal living take a different path.

The book looks deeply into the mindsets, practices and wisdom of First Nation peoples across multiple disciplines. The goals of the book are to (a) Increase understanding of "First" ways, (b) Demonstrate how traditional Indigenous cultures foster wisdom, morality and flourishing, (c) Note commonalities among different Indigenous societies in fostering these outcomes, and (d) Identify synergistic approaches to shifting human imagination towards First ways. We expect that the book will help readers envision ways to move toward integrating helpful modern advances with First ways. We all need to figure out how to blend these perspectives into a new encompassing viewpoint, one where the greater community of life (diverse human and other-than-human entities) is included in conceptions of wellbeing and in practices for sustainability that lead to global flourishing. All efforts need to help stop the continued genocide and culturecide of Indigenous People while assisting decolonizing barriers to First Nation sovereignty. Endeavors to improve community life need to be a collaboration with Indigenous wisdom keepers who can help guide local policy considerations.

Humanity's Past

There are five sections to the book. The first section focuses on humanity's past. For 99% of human genus history around the world, small-band hunter-gatherer societies lived sustainably or perished.[55] Documentation of the Kung and Australian Aborigines indicate today's communities continue social practices that are at least 40,000–150,000 years old.[56]

In Chapter 2, "The Deep Past: What Can Ancient Hunter-Gatherers Tell Us About Human Nature?" Penny Spikins provides an account of the archaeology of caring for the vulnerable, demonstrating how it reaches into the far distant past, where "societies had developed a unique kind of collaboration, involving an interdependence within groups, and between people and nature." She begins with a personal story of time spent with her grandfather, whose vision of his environment echoed that of many Indigenous populations and inspired respect for humanity's shared distant ancestors.

In Chapter 3, "Indigenous Bodies, Civilized Selves, and the Escape from the Earth," Eugene Halton maintains that history can be understood as involving a problematic interplay between the long-term legacy of human evolution, still tempered into the human body today, and the shorter-term heritage of civilization from its beginnings to the present. Each of us lives in a tension between our Indigenous bodies and our civilized selves, between the philosophy of the earth and what he characterizes as "the philosophy of escape from the earth." The standard story of civilization is that of linear upward progress, a story that he contests with an alternative philosophy of history, picturing history instead as a set of contracting concentric circles. Civilization marked a radical shift in the human relation to habitat and the development of anthropocentric mind—contracting from long term evolutionary attunement with the informing properties of wild nature to a human centered outlook progressively dependent on constructions of domesticated settlement. In the modern era the process of contraction continued through the elevation of the machine and mechanico-centric mind, involving progress in precision, coupled with a progressive regression from a sustainable worldview. Far from controlling nature, humans have been consuming it in an unsustainable Malthusian-like trajectory whose limits are being reached in our time.

In her chapter, "'Woman Is the Mother of All': Rising from the Earth," Barbara Alice Mann offers contradistinctions between the economy of invasion, from which statist capitalism grew, and the economy of the gift, from which matriarchal decentralized "twinship" forms of confederation and distribution grew. With the scholarly frankness that characterizes her work, we learn that the dismissal of our Indigenous matriarchal gift economies, whether by academics, policy-makers, or modern feminists, continues to be dangerously and erroneously predominant. Whereas Western, statist capitalism is often presented as the only modern mode of governance, in fact, the decentralized governance of the gift economy has long existed and provides a proven, and democratic, alternative.

In Chapter 5, Darcia Narvaez addresses the "Original Practices for Becoming and Being Human." She explores the source of the two worldviews (the original Indigenous and the dominant), describing how they are initiated from childhood experiences. Humans are more malleable than any other animal, like fetuses for 18 months postnatally, requiring intensive, communal, and lengthy care. She describes the shifted baseline for nurturing in early life that governs childhoods in industrialized nations. Instead of providing the humanity's evolved nest (soothing perinatal experiences, extensive on-request breastfeeding, extensive affection and responsive care, multiple adult caregivers, self-directed play through childhood, and positive social support), the industrialized world provides minimal care to babies and young children which leads to suboptimal development of neurobiological systems that govern personality and capacities for life. The dysregulation that results explains in part the soulessness of European invaders that shocked native peoples and the competitive detachment that characterizes industrialized culture.

Doing Science in Relation to Nature

The second section of the book addresses ways of doing science and relating to nature. There is a sharp contrast between contemporary Western science, though not Western science in its early days, and Indigenous science. In the early days of Western science, knowledge rested on observation and deduction, much like Indigenous science. But now, modern science uses methods of experimentation and measurement, downgrading the value of observation and deduction. The detachment promoted by contemporary Western science advocates may have a great deal to do with the ongoing destruction of the natural landscape, biodiversity, and ecological integrity. As Rupert Sheldrake commented: "knowledge gained through experience or plants and animals is not an inferior substitute for proper scientific knowledge: it is the real thing. Direct experience is the only way to build up an understanding that is not only intellectual but intuitive and practical, involving the sense and the heart as well as the rational mind."[57] This is the Indigenous view.

In Chapter 6, "Science Education as Moral Education," Greg Cajete presents Indigenous, or Native, science. He continues his preeminent work in bridging Indigenous and Western sciences with an explanation about the intersection of plant and human nature that has been a foundation for sustainable living for most of our history on this planet. Plant-human relationships may be the core dynamic in Native thinking as it relates to healthful living. Cajete's fascinating stories about the spiritual ecology that connects different cultures to certain plants can inspire Natives and non-Natives alike to "once again apply our collective and historical ability to integrate differences." So doing can bring back a flowing, ethical balance between individuals, community, and the environment of which we are all part.

In Chapter 7, keeping with his prolific legacy of research and publications about American Indian contributions to democracy, truthful communication, gender fluidity, and ecological sustainability, Bruce E. Johansen pulls no punches. In "Mother Earth vs. Mother Lode: Native Environmental Ethos, Sustainability, and Human Survival" he shares pieces of ecological wisdom from a variety of Indigenous wisdom keepers, he brings us to the dawn of the Anthropocene and details frankly and lucidly the catastrophe unfolding around the world. He explains why a "mother-lode mentality" and consequential forms of capitalism and colonization are largely responsible human-caused climate change. These in turn result from having long dismissed (or worse) Native paradigms. For example, he explains how the Indigenous consideration of "the 7th generation" is much more in alignment with the problem of "thermal inertia" than the usual environmental education put forth by schools and governments.

Ways of Being Human

The third section of the book focuses on ways of being a human being. One striking contrast between Native and Western people is the way in which senses are developed and for what end. Western people emphasize sight with a degradation of other senses.[58] Living with nature and other-than-humans is fundamental for native peoples. As a result, all senses are highly developed to facilitate respectful, sustainable living in and with nature. Some Europeans and early Americans were fascinated by native peoples' apparent instinctual understanding for animals, plants, and weather patterns—"as if all these life forms somehow engaged in transparent telecommunication."[59] First-Nation peoples understand that just as words are only the tip of the iceberg of what there is to understand, so too manifest reality is only the tip of the iceberg of a large invisible world of meaning. "Everything animate and inanimate has within it a spirit dimension and communicates *in that dimension* to those who can listen."[60] Rather than communication being about conveying information, a secondary concern, it is an

> extended "communion," the natural feeling of oneness with the many kingdoms of nature. Since speaking, singing, dancing, and chanting were all considered sacred, it

was the tone, atmosphere, and specific spirit … that rendered expression meaningful. To study a song apart from the universal spirit expressing it would be like trying to feel a heartbeat by weighing and measuring an isolated, dead heart."[61]

Healing and spirituality are not separated in First-Nation cultures. It must be pointed out that though there are many shared practices (e.g., vision quest), "every Indian [or Indigenous] tribe has a spiritual heritage that distinguishes them from all other people."[62]

In his chapter, "The Structure and Transmission of Tlingit Principles and Practices of Sustainable Wisdom," anthropologist Steve J. Langdon describes how traditional Tlingit wisdom fosters a "continuous circulation of respect in thought, speech and deed" that lead to "relational sustainability" for all living and nonliving forms, with special focus on relationships with salmon. Tlingit are guided in their relations with salmon by a mythic charter/covenant that informs them that salmon are "persons" like humans and illuminates how people are to treat the salmon in order for salmon to return in the future. His anthropological research reveals how intergenerational wisdom about life and the cosmos has been successful for thousands of years and continues today in spite of increasing external assaults. Although Tlingit "obligatory reciprocity" and pedagogical practices are unique to Tlingit culture and geography, Dr. Langdon presents them as powerful lessons for all who are striving toward ecological balance in the world, especially for educators and parents looking for ways to teach moral obligations and a deep understanding of the consequences for not heeding them.

In Chapter 9, "Indigenous Spirituality: A Matter of Significance," Four Arrows courageously speaks truth to power and describes the heart of a sustainable spirituality. Fearless engagement with the truth has been a characteristic of Indigenous societies, but not of dominant civilized cultures that instead use deception as a matter of course and are fearful of truth. Four Arrows describes his experience as a Standing Rock Water Protector (against the Dakota Access Pipeline) where he and others stood rooted in the truth of a sacred earth against the forces of deceit and destruction. He also describes his habitual practice for maintaining a truthful orientation to the earth, practices that ultimately open our minds to the relatedness of all and the sacredness of all relationships, revealing that our true natures can only be found on a path guided by a community led by fearless engagement in truthfulness.

In Chapter 10, "Listening to the Trees," Tom McCallum seamlessly weaves the story of his transformative childhood vision into the great vision of balance in the spiral of life and the ceremony of the Sundance. He also gives an inside view from the original instructions in its visionary language, and provides examples of how to understand these ideas from outside of science, with evidence from *The Secret Life of Trees*, fractals, and other sources of data. The ideas of transformative spiritual energy as embedded in original instructions and encountered in vision and ceremony are not easy for the modern mind to accept or even understand. Yet this chapter does a wonderful job of stating the ideas themselves firsthand as well as providing contexts for them.

Integrated Futures

Section four anticipates integrated futures. When Europeans came, American Indians were "wrenched from a free life where the natural order had to be understood and obeyed, [and then] confined within a foreign educational system where memorization and recital substitute for learning and knowledge."[63] Critical to living on the earth is knowing its systems and their interactions. Although general knowledge about systems and interactions can be useful, true understanding is localized. Traditional Ecological Knowledge (TEK), that which Indigenous peoples preserve, is required understanding for respectful, sustainable behavior. TEK comes from immersion in a landscape, learning from observation and experience, building deep tacit (nonverbal) knowledge from which intuitions about right actions arise.[64] Moves required to integrate First and modern ways include restoring a land ethic as well as restoring capacities for receptivity to and communications with other-than-humans and the respect that goes with both.

In Chapter 11, Jon Young brings a wide range of profound insights both into nature connection and into causes of our current state of disconnect. In his chapter, "Nature Sense to Innate Wisdom: Effective Connection Modeling & Regenerating Human Beings," he shows how, despite good intentions, our educational, recreational, and wellness-focused institutions ignore practiced immersion in nature—something foundational to human nature for both learning and development. Young notes the ways it affects how children are raised. Based on his decades of experience as a master wildlife tracker and mentor for nature immersion programs, he provides a number of concrete and already proven ways to remedy the state of disconnect in individual people, as well as the in culture at large.

In Chapter 12, Sandra Waddock applies the notion of shamanism to modern life in "Modern (Intellectual) Shamans and Wisdom for Sustainability." She encourages us to exert our own shamanic power to help heal the troubles that the planet is facing. For scholars, this means resolutely orienting our intellectual work towards healing, connecting, and sense-making in the service of a better world. By considering the Earth itself as a living entity, we would accord dignity—inherent worth—and wellbeing not only to humans, but also to all other living beings and all aspects of nature. Such a shamanic approach radically changes our scholarly approach to business and other activities. It would also help companies and economies to become truly sustainable while still enabling them to be successful.

Finally, in "For a Tattered Planet: Art and Tribal Continuance," poet Kim Blaeser provides us with deep hope as she tells us the story of why Native people create art and further, why she creates art through her words for her people—for all people. Blaeser writes, "Native spaces still exist in our everyday landscapes," and goes on to explain and show "how readily we can re-inhabit them." This idea that all is not lost, but rather that we can re-inhabit the things that are special, important, and spiritual is fundamental to our human experience; we must believe that we have the opportunity and the ability to re-claim and re-inhabit. As Blaeser shows in this work Native artists are revered because, as Paul Chaat Smith says, they are asked "to lead the revolution." Blaeser discusses this revolution and shows how it looks in the arms of beauty through her picto-poems that she includes in this chapter.

Conclusion

If there could be one singular message emerging from the chapters of this volume, it would be the restoration of relationship with the living earth. The current imbalance of the earth is not due to natural cataclysm or catastrophe, but to human misdirection in ways of living that are out of balance with the laws of nature. Regaining our own balance is key to human flourishing as it has been for most of human history. Our greatest challenge is to realize that technology alone will not help—not without a change in the orientation that has brought us to the brink of a mass extinction. We can re-adopt the mindset that exists in the worldview and cultures of Indigenous hunter-gatherer of the past and of those managing against all odds to hold on to them today. It is our hope that the offerings in this book provide inspiration and guidance for finding ways to weave this Indigenous wisdom into our own lives and cultures before it is too late. Avoiding romantic notions about some idealized past while borrowing from the deep understandings that kept us in balance for most of our history on this planet, we can learn important lessons from Indigenous peoples while doing what we can to help prevent contemporary Indigenous cultures from vanishing.

Instead of plans to escape from Mother Earth, our scientific endeavors must do everything possible to put us back into harmony with her. This calls for engaging all of our senses, leading with our hearts and spirits, as well as the rational mind. To live sustainably and wisely means to live in participation with the natural world, fostering biophilia through daily practices of respect and understanding of the natural world, particularly one's own landscape. This comes about through "becoming open to the natural world with all of one's senses, body, mind, and spirit." [65] When childhood has undermined this openness, the adolescent or adult must deliberately work to repair and relearn how to relax into immersion in the natural world where "receptivity, creativity, perception, and imagination are integrated through participation with nature." [66]

Communication is more than conveying information; rather, as an extended communion, it draws on the natural feeling of oneness with the many "kingdoms" of nature. Listening to the spirit dimension in all life forms sets up the way to truthfulness and sustainable wisdom. With the information in this book, we can redirect communal life that is currently dominated by unchecked egos and invasive powerful institutions and practices. We can find ways to create gift economies that are decentralized and inclusive. By according inherent worth to all living beings and all aspects of nature, we will radically move all towards true sustainability. In fact, although it seems inconspicuous, this is already beginning to happen. A return to sustainable wisdom is emerging if we give attention and respect to our interconnectedness. Can we hear the trees singing?

Notes

1. Frederick W. Turner, *Beyond Geography: The Western Spirit against the Wilderness* (New Brunswick, NJ: Rutgers University Press, 1994).
2. Turner, *Beyond Geography*, 260.
3. Aaron Carapella, "Tribal Nations Maps," http://www.tribalnationsmaps.com/.
4. Benjamin Madley, *An American Genocide: The United States and the California Indian Catastrophe, 1846–1873* (New Haven, CT: Yale University Press, 2016).
5. Elizabeth Kolbert, *The Sixth Extinction: An Unnatural History* (New York, NY: Henry Holt, 2014).
6. Kolbert, *The Sixth Extinction*; Stephen R. Carpenter, *Ecosystems and Human Well-Being: Scenarios: Findings of the Scenarios Working Group, Millennium Ecosystem Assessment* (Washington, D.C.: Island Press, 2005).
7. Intergovernmental Panel on Climate Change, *Climate change 2013: The physical science basis. Working Group I Contribution to the 5th Assessment Report of the Intergovernmental Panel on Climate Change: Changes to the Underlying Scientific/Technical Assessment to ensure consistency with the approved Summary for Policymakers* (Geneva, Switzerland: United Nations, 2013). Will Steffen el al., "Trajectories of the earth system in the Anthropocene," *Proceedings of the National Academy of Sciences*, 115, no. 33 (2018): 8252–8259.
8. David Korten, *Change the Story, Change the Future* (Oakland, CA: Berrett-Koehler, 2015).
9. Peter Capossela, "Impacts of the Army Corps of Engineer's Pick-Sloan Program on the Indian Tribes of the Missouri River Basin," *Journal of Environmental Law and Litigation* 30, no. 1 (2015): 143–217.

10. Pope Francis, *Laudato Si': On Care for Our Common Home* [Encyclical letter] (Huntington, IN: Our Sunday Visitor, 2015), paragraphs 146 and 147.
11. Intergovernmental Panel on Climate Change, 2013; Carpenter, 2005.
12. Anne Case and Angus Deaton, "Mortality and Morbidity in the 21st Century," *Brookings* (23 March 2017) https://www.brookings.edu/bpea-articles/mortality-and-morbidity-in-the-21st-century/; "Rising Morbidity and Mortality in Midlife among White Non- Hispanic Americans in the 21st Century," *Proceedings of the National Academy of Sciences of the United States of America* 112, no. 49 (2015); Steven H. Woolf and Laudan Y. Aron, *U.S. Health in International Perspective: Shorter Lives, Poorer Health* (Washington, D.C.: The National Academies Press, 2013); Organisation for Economic Cooperation and Development, *Doing Better for Children* (Paris: OECD, 2009); Jiaquan Xu et al., "Mortality in the United States, 2015," *National Center for Health Statistics*, no. 267 (2016).
13. Clark Spencer Larsen, "The Agricultural Revolution as Environmental Catastrophe: Implications for Health and Lifestyle in the Holocene," *Quaternary International* 150, no. 1 (2006): 12–20.
14. Alan Eliare, *From the Heart of the World: The Elder Brother's Warning* [film] (London, UK: British Broadcasting Corporation, 1990), quoted in Tom Cooper, *A Time before Deception: Truth in Communication, Culture, and Ethics* (Santa Fe, NM: Clear Light, 1998), 98.
15. Ibid.
16. J. Baird Callicott, *In Defense of the Land Ethic: Essays in Environmental Philosophy* (Albany, NY: State University of New York Press, 1989), 90.
17. Kolbert, *The Sixth Extinction*.
18. Derrick Jensen, *The Myth of Human Supremacy* (New York, NY: Seven Stories Press, 2016).
19. Jason Moore, "The Capitalocene, Part I: On the Nature and Origins of Our Ecological Crisis," *The Journal of Peasant Studies* 44, no. 3 (2017): 594–630.
20. Robert Redfield, *The Primitive World and Its Transformations* (Ithaca, NY: Cornell University Press, 1953); Four Arrows, *Point of Departure: Returning to Our More Authentic, Worldview for Education and Survival* (Charlotte, NC: Information Age, 2016); Four Arrows and Darcia Narvaez, "Reclaiming Our Indigenous Worldview," in *Working for Social Justice inside and Outside the Classroom: A Community of Students, Teachers, Researchers, and Activists*, ed. Nancy E. McCrary and E. Wayne Ross (New York: Peter Lang, 2016). 93–112
21. Robert Redfield, *The Primitive World and Its Transformations*.
22. Philip Descola, *Beyond Nature and Culture*, trans. Janet Lloyd, (Chicago, IL: University of Chicago Press, 2013).
23. Ibid.
24. Lewis Mumford, *The Myth of the Machine*, 1st ed. (New York, NY: Harcourt, Brace & World, 1967); Markus Christen, Darcia Narvaez, and Eveline Gutzwiller, "Comparing and Integrating Biological and Cultural Moral Progress," *Ethical Theory and Moral Practice*, 20, no. 1 (2017): 55.
25. Carolyn Merchant, *The Death of Nature: Women, Ecology, and the Scientific Revolution*, 1st ed. (San Francisco, CA: Harper & Row, 1980).
26. Bruno Latour, *An Inquiry into Modes of Existence: An Anthropology of the Moderns* (Cambridge, MA: Harvard University Press, 2013); Mumford, *The Myth of the Machine*.
27. Thomas Berry, *The Great Work: Our Way into the Future* (New York, NY: Bell Tower, 1999); *The Dream of the Earth* (San Francisco, CA: Sierra Club Books, 1990).

28. Marshall Sahlins, *The Western Illusion of Human Nature: With Reflections on the Long History of Hierarchy, Equality and the Sublimation of Anarchy in the West, and Comparative Notes on Other Conceptions of the Human Condition* (Chicago, IL: Prickly Pear Paradigm Press, 2008).

29. Christen, Narvaez, and Gutzwiller, "Comparing and Integrating Biological and Cultural Moral Progress."

30. Paul Shepard, *Nature and Madness*, ed. C. L. Rawlins (San Francisco, CA: Sierra Club Books, 1982; Athens, GA: University of Georgia Press, 1998); *Coming Home to the Pleistocene*, ed. Florence R. Shepard (Washington, D.C.: Island Press, 1998).

31. Cooper, *A Time before Deception*; Tim Ingold, "On the Social Relations of the Hunter-Gatherer Band," in *The Cambridge Encyclopedia of Hunters and Gatherers*, ed. Richard Lee and Richard Daly (New York, NY: Cambridge University Press, 2005).

32. Douglas P. Fry, *The Human Potential for Peace: An Anthropological Challenge to Assumptions About War and Violence* (New York: Oxford University Press, 2006).

33. E.g. Cooper, *A Time before Deception*; Vine Deloria, *The World We Used to Live In: Remembering the Powers of the Medicine Men* (Golden, CO: Fulcrum, 2006).

34. Fry, *The Human Potential for Peace*; Ingold, "On the Social Relations of the Hunter-Gatherer Band."

35. Darcia Narvaez, *Neurobiology and the Development of Human Morality: Evolution, Culture and Wisdom* (New York, NY: W. W. Norton, 2014).

36. Shepard, *Coming Home to the Pleistocene*.

37. Darcia Narvaez, *Ancestral Landscapes in Human Evolution: Culture, Childrearing and Social Wellbeing* (Oxford, UK: Oxford University Press, 2014); Jaak Panksepp, "The Long-Term Psychobiological Consequences of Infant Emotions: Prescriptions for the Twenty-First Century," *Infant Mental Health Journal* 22, no. 1–2 (2001):132–73; Eugene Halton, "Eden Inverted: On the Wild Self and the Contraction of Consciousness," *The Trumpeter* 23, no. 3 (2007); James W. Prescott, "Origins of Love & Violence and the Developing Human Brain," *Pre- and Perinatal Psychology Journal* 10, no. 3 (1996): 143–88; Allan N. Schore, "The Effects of Early Relational Trauma on Right Brain Development, Affect Regulation, and Infant Mental Health," *Infant Mental Health Journal* 22, no. 1–2 (2001): 201–69; Silvan S. Tomkins, "Affect and the Psychology of Knowledge," in *Affect, Cognition, and Personality*, ed. Carroll E. Izard and Silvan S. Tomkins (New York, NY: Springer, 1965).

38. Darcia Narvaez and Tracy Gleason, "Developmental Optimization," in *Evolution, Early Experience and Human Development: From Research to Practice and Policy*, ed. Darcia Narvaez, et al. (New York, NY: Oxford University Press, 2013), 307–325.

39. Darcia Narvaez, "The Neurobiology of Moral Sensitivity: Evolution, Epigenetics and Early Experience," in *The Art of Morality: Developing Moral Sensitivity Across the Curriculum*, ed. Deborah Mowrer and Peggy Vandenberg (New York, NY: Routledge, 2015), 19–42.

40. Darcia Narvaez, "Baselines for Virtue," in *Developing the Virtues: Integrating Perspectives*, ed. Julia Annas, Darcia Narvaez, and Nancy Snow (New York, NY: Oxford University Press, 2016), 14–33.

41. Charles Derber, *Sociopathic Society: A People's Sociology of the United States* (Boulder: Paradigm, 2013).

42. Turner, *Beyond Geography*.

43. Jean M. O'Brien, *Firsting and Lasting: Writing Indians out of Existence in New England* (Minneapolis, MN: University of Minnesota Press, 2010).

44. Roxanne Dunbar-Ortiz, *An Indigenous People's History of the United States* (Boston, MA: Beacon Press, 2014).

45. Darcia Narvaez and David Witherington, "Getting to baselines for human nature, development and wellbeing." *Archives of Scientific Psychology*, 6, no. 1 (2018): 205; Narvaez, *Neurobiology and the Development of Human Morality.*

46. Judith L. Bronstein, ed., *Mutualism* (New York: Oxford University Press, 2015); Rob Dunn, *The Wild Life of Our Bodies: Predators, Parasites, and Partners that Shape Who We Are Today* (New York, NY: Harper, 2011); Peter Wohlleben, *The Hidden Life of Trees: What They Feel, How They Communicate*, trans. Jane Billinghurst (Vancouver, BC: Greystone Books, 2016).

47. Elise Amel, Christie Manning, Britain Scott and Susan Koger, "Beyond the roots of human inaction: Fostering collective effort toward ecosystem conservation," *Science*, 356 (2017): 275–279; Douglas L. Medin and Megan Bang, *Who's Asking?: Native Science, Western Science, and Science Education* (Cambridge, MA: MIT Press, 2014).

48. Edwina Pio et al., "Pipeline to the Future: Seeking Wisdom in Indigenous, Eastern, and Western Traditions," in *Handbook of Faith and Spirituality in the Workplace: Emerging Research and Practice*, ed. Judi Neal (New York, NY: Springer, 2013), 195–219.

49. Note: We address this group of peoples with various terms throughout the book, including Indigenous; Aboriginal; First Nations; Original People; etc. We note that Lakota activist, Russell Means prefers "American Indian" to "Native American" as it refers specifically and only to the people originally colonized by the Europeans. When possible, it is best to refer to the actual name of the tribe. See http://www.pbs.org/wgbh/roadshow/fts/bismarck_200504A16.html);

50. E.g., Robin Wall Kimmerer, *Braiding Sweetgrass: Indigenous Wisdom, Scientific Knowledge and the Teachings of Plants*, 1st ed. (Minneapolis, MN: Milkweed Editions, 2013).

51. Dennis Martinez et al., "Restoring Indigenous History and Culture to Nature," in *Original Instructions*, ed. Melissa K. Nelson (Rochester, VT: Bear & Company, 2008), 88–115.

52. Kimmerer, *Braiding Sweetgrass.*

53. Shepard, *Coming Home to the Pleistocene.*

54. Jerome S. Bernstein, *Living in the Borderland: The Evolution of Consciousness and the Challenge of Healing Trauma* (New York, NY: Routledge, 2005), 8.

55. Douglas Fry, ed., *War, Peace and Human Nature* (New York, NY: Oxford University Press, 2013).

56. Michael Balter, "Ice Age Tools Hint at 40,000 Years of Bushman Culture," *Science* 337, no. 6094 (2012): 512; Robert Lawlor, *Voices of the First Day: Awakening in the Aboriginal Dreamtime* (Rochester, VT: Inner Traditions International, 1991); James Suzman, *Affluence without abundance: The disappearing world of the Bushmen* (New York: Bloomsbury, 2017).

57. Rupert Sheldrake, *The Rebirth of Nature: The Greening of Science and God* (New York, NY: Bantam Books, 1991), 213.

58. Harold A. Innis, *The Bias of Communication* (Toronto, ON: University of Toronto Press, 1951); Marshall McLuhan, *Understanding Media* (New York, NY: McGraw-Hill, 1964).

59. Cooper, *A Time before Deception*, 26. See also Laura D. Walls, *Henry David Thoreau: A life* (Chicago, IL: University of Chicago Press, 2017) and Descola, *Beyond Nature and Culture.*

60. Bernstein, *Living in the Borderland.*
61. Cooper, *A Time before Deception*, 16.
62. Deloria, *The World We Used to Live In*, xxiii.
63. Ibid., xviii.
64. Michael Polanyi, *Personal Knowledge: Towards a Post-Critical Philosophy* (Chicago, IL: University of Chicago Press, 1958).
65. Greogry Cajete, *Native Science: Natural Laws of Interdependence* (Santa Fe, NM: Clear Light, 2000), 21.
66. Ibid., 26.

References

Amel, Elise, Christie Manning, Britain Scott and Susan Koger, "Beyond the Roots of Human Inaction: Fostering Collective Effort Toward Ecosystem Conservation," *Science*, 356 (2017): 275–9.

Balter, Michael. "Ice Age Tools Hint at 40,000 Years of Bushman Culture." *Science* 337, no. 6094 (2012): 512.

Bernstein, Jerome S. *Living in the Borderland: The Evolution of Consciousness and the Challenge of Healing Trauma.* New York, NY: Routledge, 2005.

Berry, Thomas. *The Dream of the Earth.* San Francisco, CA: Sierra Club Books, 1990.

———. *The Great Work: Our Way into the Future.* New York, NY: Bell Tower, 1999.

Bronstein, Judith L., ed. *Mutualism.* New York, NY: Oxford University Press, 2015.

Cajete, Gregory. *Native Science: Natural Laws of Interdependence.* Santa Fe, NM: Clear Light, 2000.

Callicott, J. Baird. *In Defense of the Land Ethic: Essays in Environmental Philosophy.* Albany, NY: State University of New York Press, 1989.

Capossela, Peter. "Impacts of the Army Corps of Engineer's Pick-Sloan Program on the Indian Tribes of the Missouri River Basin." *Journal of Environmental Law and Litigation* 30, no. 1 (2015): 143–217.

Carapella, Aaron. "Tribal Nations Maps." http://www.tribalnationsmaps.com/.

Carpenter, Stephen R. *Ecosystems and Human Well-Being: Scenarios: Findings of the Scenarios Working Group, Millennium Ecosystem Assessment.* Washington, D.C.: Island Press, 2005.

Case, Anne, and Angus Deaton. "Mortality and Morbidity in the 21st Century." *Brookings*, 23 March 2017. https://www.brookings.edu/bpea-articles/mortality-and-morbidity-in-the-21st-century/.

———. "Rising Morbidity and Mortality in Midlife among White Non- Hispanic Americans in the 21st Century." *Proceedings of the National Academy of Sciences of the United States of America* 112, no. 49 (2015): 15078–83.

Christen, Markus, Darcia Narvaez, and Eveline Gutzwiller. "Comparing and Integrating Biological and Cultural Moral Progress." *Ethical Theory and Moral Practice* 20, no. 1 (2017): 55. doi:10.1007/s10677-016-9773-y.

Cooper, Tom. *A Time before Deception: Truth in Communication, Culture, and Ethics*. Santa Fe, NM: Clear Light, 1998.

Deloria, Vine. *The World We Used to Live In: Remembering the Powers of the Medicine Men*. Golden, CO: Fulcrum, 2006.

Derber, Charles. *Sociopathic Society: A People's Sociology of the United States*. Boulder: Paradigm, 2013.

Descola, Phillip. *Beyond Nature and Culture*. Translated by Janet Lloyd. Chicago, IL: University of Chicago Press, 2013.

Dunn, Rob. *The Wild Life of Our Bodies: Predators, Parasites, and Partners that Shape Who We Are Today*. New York, NY: Harper, 2011.

Dunbar-Ortiz, Roxanne. *An Indigenous People's History of the United States*. Boston, MA: Beacon Press, 2014.

Eliare, Alan. *From the Heart of the World: The Elder Brother's Warning* [film]. London, UK: British Broadcasting Corporation, 1990. Quoted in Tom Cooper, *A Time before Deception: Truth in Communication, Culture, and Ethics*. Santa Fe, NM: Clear Light, 1998.

Four Arrows (aka Don Trent Jacobs). *Point of Departure: Returning to Our More Authentic, Worldview for Education and Survival*. Charlotte, NC: Information Age, 2016.

_____ and Darcia Narvaez. "Reclaiming Our Indigenous Worldview." In *Working for Social Justice Inside and Outside the Classroom: A Community of Students, Teachers, Researchers, and Activists*, edited by Nancy E. McCrary and E. Wayne Ross, 93–112. New York: Peter Lang, 2016.

Francis, Pope. *Laudato Si': On Care for Our Common Home* [Encyclical letter]. Huntington, IN: Our Sunday Visitor, 2015.

Fry, Douglas P. *The Human Potential for Peace: An Anthropological Challenge to Assumptions About War and Violence*. New York, NY: Oxford University Press, 2006.

———, ed. *War, Peace and Human Nature*. New York, NY: Oxford University Press, 2013.

Halton, Eugene. "Eden Inverted: On the Wild Self and the Contraction of Consciousness." *The Trumpeter* 23, no. 3 (2007): 45–77.

Ingold, Tim. "On the Social Relations of the Hunter-Gatherer Band." In *The Cambridge Encyclopedia of Hunters and Gatherers*, edited by Richard Lee and Richard Daly, 399–410. New York, NY: Cambridge University Press, 2005.

Innis, Harold A. *The Bias of Communication*. Toronto, ON: University of Toronto Press, 1951.

Intergovernmental Panel on Climate Change. *Climate change 2013: The physical science basis. Working Group I Contribution to the 5th Assessment Report of the Intergovernmental Panel on Climate Change: Changes to the Underlying Scientific/Technical Assessment to ensure consistency with the approved Summary for Policymakers*. IPCC-XXXVI/Doc.4, Agenda Item 3. Geneva, Switzerland: United Nations, 27 September 2013.

Jensen, Derrick. *The Myth of Human Supremacy*. New York, NY: Seven Stories Press, 2016.

Kimmerer, Robin Wall. *Braiding Sweetgrass: Indigenous Wisdom, Scientific Knowledge and the Teachings of Plants* (1st ed.). Minneapolis, MN: Milkweed Editions, 2013.

Kolbert, Elizabeth. *The Sixth Extinction: An Unnatural History* (1st ed.). New York, NY: Henry Holtand Company, 2015.

Korten, David. *Change the Story, Change the Future.*Oakland, CA: Berrett-Koehler, 2015.

Larsen, Clark Spencer. "The Agricultural Revolution as Environmental Catastrophe: Implications for Health and Lifestyle in the Holocene." *Quaternary International* 150, no. 1 (2006): 12–20.

Latour, Bruno. *An Inquiry into Modes of Existence: An Anthropology of the Moderns.* Cambridge, MA: Harvard University Press, 2013.

Lawlor, Robert. *Voices of the First Day: Awakening in the Aboriginal Dreamtime.* Rochester, VT: Inner Traditions International, 1991.

Madley, Benjamin. *An American Genocide: The United States and the California Indian Catastrophe, 1846–1873.* New Haven, CT: Yale University Press, 2016.

Martinez, Dennis, Enrique Salmon, and Melissa K. Nelson. "Restoring Indigenous History and Culture to Nature." In *Original Instructions*, edited by Melissa K. Nelson, 88–115. Rochester, VT: Bear& Company, 2008.

McLuhan, Marshall. *Understanding media.* New York, NY: McGraw-Hill, 1964.

Medin, Douglas L., and Megan Bang. *Who's Asking? Native Science, Western Science, and Science Education.* Cambridge, MA: MIT Press, 2014.

Merchant, Carolyn. *The Death of Nature: Women, Ecology, and the Scientific Revolution* (1st ed.). San Francisco, CA: Harper & Row, 1980.

Moore, Jason. "The Capitalocene, Part I: On the Nature and Origins of Our Ecological Crisis." *The Journal of Peasant Studies* 44, no. 3 (2017): 594–630.

Mumford, Lewis. *The Myth of the Machine* (1st ed.). New York, NY: Harcourt, Brace & World, 1967.

Narvaez, Darcia. *Ancestral Landscapes in Human Evolution: Culture, Childrearing and Social Wellbeing.* Oxford, UK: Oxford University Press, 2014.

———. "Baselines for Virtue." In *Developing the Virtues: Integrating Perspectives*, edited by Julia Annas, Darcia Narvaez and Nancy Snow, 14–33. New York, NY: Oxford University Press, 2016.

———. *Neurobiology and the Development of Human Morality: Evolution, Culture and Wisdom.* New York, NY: W. W. Norton, 2014.

———. "The Neurobiology of Moral Sensitivity: Evolution, Epigenetics and Early Experience." In *The Art of Morality: Developing Moral Sensitivity Across the Curriculum*, edited by Deborah Mowrer and Peggy Vandenberg, 19–42. New York, NY: Routledge, 2015.

——— and Tracy Gleason. "Developmental Optimization." In *Evolution, Early Experience and Human Development: From Research to Practice and Policy*, edited by Darcia Narvaez, Tracy Gleason, Jaak Panksepp, and Allan Schore, 307–325. New York, NY: Oxford University Press, 2013.

——— and David Witherington, "Getting to baselines for human nature, development and wellbeing," *Archives of Scientific Psychology*, 6 (2018): 205–13.

O'Brien, Jean M. *Firsting and Lasting: Writing Indians out of Existence in New England.* Minneapolis, MN: University of Minnesota Press, 2010.

Organisation for Economic Cooperation and Development. *Doing Better for Children.* Paris: OECD, 2009.

Panksepp, Jaak. "The Long-Term Psychobiological Consequences of Infant Emotions: Prescriptions for the Twenty-First Century." *Infant Mental Health Journal* 22, no. 1–2 (2001): 132–73.

Pio, Edwina, Sandra Waddock, Mzamo Mangaliso, Malcolm McIntosh, Chellie Spiller, Hiroshi Takeda, Joe Gladstone, Marcus Ho, and Jawad Syed. "Pipeline to the Future: Seeking Wisdom in Indigenous, Eastern, and Western Traditions." In *Handbook of Faith and Spirituality in the Workplace: Emerging Research and Practice*, edited by Judi Neal, 195–219. New York, NY: Springer, 2013.

Polanyi, Michael. *Personal Knowledge: Towards a Post-Critical Philosophy*. Chicago, IL: University of Chicago Press, 1958.

Prescott, James W. "Origins of Love & Violence and the Developing Human Brain." *Pre- and Perinatal Psychology Journal* 10, no. 3 (1996): 143–88.

Redfield, Robert. *The Primitive World and Its Transformations*. Ithaca, NY: Cornell University Press, 1953.

Sahlins, Marshall. *The Western Illusion of Human Nature: With Reflections on the Long History of Hierarchy, Equality and the Sublimation of Anarchy in the West, and Comparative Notes on Other Conceptions of the Human Condition*. Chicago, IL: Prickly Pear Paradigm Press, 2008.

Schore, Allan N. "The Effects of Early Relational Trauma on Right Brain Development, Affect Regulation, and Infant Mental Health." *Infant Mental Health Journal* 22, no. 1–2 (2001): 201–69.

Sheldrake, Rupert. *The Rebirth of Nature: The Greening of Science and God*. New York, NY: Bantam Books, 1991.

Shepard, Paul. *Coming Home to the Pleistocene*. Edited by Florence R. Shepard. Washington, D.C.: Island Press, 1998.

———. *Nature and Madness*. Edited by C. L. Rawlins. San Francisco, CA: Sierra Club Books, 1982; Athens, GA: University of Georgia Press, 1998.

Tomkins, Silvan S. "Affect and the Psychology of Knowledge." In *Affect, Cognition, and Personality*, edited by Carroll E. Izard and Silvan S. Tomkins. New York, NY: Springer, 1965.

Turner, Frederick W. *Beyond Geography: The Western Spirit Against the Wilderness*. New Brunswick, NJ: Rutgers University Press, 1994.

Walls, Laura D. *Henry David Thoreau: A Life*. Chicago, IL: University of Chicago Press, 2017.

Wohlleben, Peter. *The Hidden Life of Trees: What They Feel, How They Communicate*. Translated by Jane Billinghurst. Vancouver, BC: Greystone Books, 2016.

Woolf, Steven H., and Laudan Y. Aron. *U.S. Health in International Perspective: Shorter Lives, Poorer Health*. Washington, D.C.: The National Academies Press, 2013.

Xu, Jiaquan, Sherry L. Murphy, Kenneth D. Kochanek, and Elizabeth Arias. "Mortality in the United States, 2015." *National Center for Health Statistics*, no. 267 (2016). Accessed 23 July 2018. https://www.cdc.gov/nchs/products/databriefs/db267.htm.

Understanding Humanity's Past

CHAPTER TWO

Conversations with the Deep Past

What Can Ancient Hunter-Gatherers Tell Us about Sustainable Wisdom?

PENNY SPIKINS

Acknowledgments

I would like to thank the Sustainable Wisdom project and conference for inviting my participation, and the Templeton Trust funded "Hidden Depths: the ancestry of our most human emotions" project [grant 59475] for time to complete this chapter.

Introduction

How do we connect with the sustainable wisdom of Indigenous communities in a world so dominated by the constant materialism, competitiveness and stress of modern life? The wisdom shared by Indigenous peoples rarely fits into the frameworks of industrialized societies. It cannot entirely be analyzed or even written down, as often Indigenous wisdom is *learned through experience*. Children in Indigenous societies are not taught or told knowledge as *information*, but *experience* a fundamental interdependence both between people, and between people and nature. Babies are turned towards others, learn to share, and learn self-control through listening and seeing what happens around them. As they grow older, myths and stories told around campfires contribute to the development of wisdom about the complexity of the human condition and our conflicting emotions—jealousies, mistakes, pride and courage—that encourages not *a fact-based knowledge* about how human minds work but a *contemplative wisdom* based on complex feelings. Learning is through exploration rather than being told. This leads to the kind of necessary practical understanding of the social and natural environment that is more about personal engagement than objectified knowledge.[1] To understand this practical wisdom from an outside view is not straightforward. As so many myths and tales from around the world tell us, we can lose some of the most important things in life when we try to control them, and if we grasp at turning Indigenous wisdom into analytical facts we may lose an important part of that wisdom itself.

However, our need to connect with Indigenous wisdom to help us to navigate a path to a sustainable future is becoming ever greater as we face problems such as climate change, inequality, and high rates of mental suffering. Yet we lack opportunities to sit and listen to stories of such wisdom as it was meant to be experienced—from elders told around camp fires. Over half of the world's population live in cities, and in many regions Indigenous populations who lived with nature have long since disappeared. For example, at least five thousand years have passed since Indigenous peoples made a living through hunting and gathering in Europe and these ancient Indigenous peoples left little behind to tell us their stories. Nonetheless, although we must preserve our existing Indigenous cultures and learn from them, we can also learn from the deep past.

The distant past tells us something important about those elements of ourselves which are constant through time and fundamental to humanity. What we learn from reflecting on those constant elements of our deep past resonates with much of the sustainable wisdom we find in modern Indigenous communities. The deep past can also be a source of indirect experience, something which affects us emotionally, which brings us out of our own lives and gives us a sense of perspective on humanity as a whole and of our place in a wider world beyond our own existence. I'd like to introduce you to this deep past through conversations with my grandfather. He had no claim to Indigenous ancestry but his connection to nature and way of *showing rather than telling* reveals much about the Indigenous approach to learning.

Time with My Grandfather

I lived as a child with my extended family of parents and grandparents on a farm in the English Midlands. From a long line of small-scale famers no one in the family stood out, or left behind any invention, product, claim to fame, or even any of their own written records. As small-scale farmers, this is nothing unusual; they depended on producing food and lived within cycles of seasons, leaving little behind. Certainly, my earliest memories are all of being outside—collecting chestnuts, pulling up potatoes, climbing trees, wading through streams, and playing in haystacks. I ran a little "wild" at times, so I'm told, insisting on wearing trousers, refusing to wear dresses or have my hair combed, and always being muddy from playing with the dogs and the pigs. I even remember our house as if slightly from above, having spent so long sitting high up in a tree with a book that this view became the norm for me. I explored outwards from home, and found dead things (so fascinating), and most of all I spent a lot of time with my grandfather.

Like most farmers, my grandfather was a man of few words, yet he was a never-ending source of things that prompted you to think and reflect. He told many stories about Indigenous peoples, as many years before he had emigrated to Australia and worked on farms alongside Indigenous peoples. His stories about their need to journey, to "go walkabout," were part of what I was brought up with, and it seemed clear that their ways of seeing the world often resonated with his. When he would pick me up from school I remember that we often listened to a radio program about the Hopi.

There was something special about spending time with my Grandfather that I'm sure represents how people lived in the ancient past. His style of humorously and safely introducing difficult and often challenging things reminds me of how the Inuit coach their children.[2] I don't remember my grandfather ever telling me anything I *should know*, or *should understand* and there wasn't much that I *should do*. I remember when riding in the car on the way home from school listening to the radio, I ate the crust from the bread he had been asked to pick up from the bakers. He probably knew my grandmother would scold us for leaving her with a crustless loaf, but he never told me not to do it.[3]

He also told me "tall tales" like the one that featured sheep in certain Welsh mountains that had been bred to have legs shorter on one side than the other so they could walk (only one way) more easily around the mountain. Even today, I sometimes think I might know something (such as that cream is at the top of the milk because it comes last out of the cow), only to realize that this was a tale spun by my grandfather. Rules became things to explore and to question and I can't help laughing when I remember the time I insisted on wearing a pair of underpants on my head when we went shopping in the local village. Rather than being embarrassed and insisting that I take them off, my grandparents just let me, and I can remember even now the shopkeepers saying, "Now, that is a very unusual hat."

The most important "conversations" with my grandfather weren't made with words however. Most of all such "conversations" (if we can call them that) told me about the things that my grandfather saw but that others were going too fast to notice. He looked in places where few people did and saw things that many others never would. When you were walking with him he might point something out or pick something up. Sometimes it would be a plant or animal that caught his attention, but more often than not these were instead really old things that told a long story of our deep past. Once he found a mammoth tusk. I remember touching this amazing thing from a lost world. Other times it would be arrowheads, stone tools, pieces of pottery. Such forgotten things seemed to be everywhere if only you went slowly enough to let them reveal themselves to you. It was those small forgotten things of a distant human past that I chose to study. Such abandoned things seem to tell us something that echoes the wisdom of modern Indigenous peoples, yet is all too often ignored—as humans *we are fundamentally dependent on each other, and on nature.*

Archaeological Misinterpretation

All too often we assume these peoples lived harsh, competitive and individualistic lives and we make interpretations on this basis, often despite, rather than because of, archaeological evidence. In the 1940s Raymond Dart was studying the earliest remains of distant three-million-year-old human ancestors in Makapansgat Cave in South Africa. He noted that there were puncture marks on their skulls and interpreted this as evidence of violence and aggression. He believed sharpened antelope bones were associated with these skulls as weapons. He saw this supposedly earliest evidence for human violence as the element that marked our ancestors as different from other apes. However close inspection of the same evidence revealed that these early australopithecine ancestors (pre-human ancestral populations) were far from the innately violent and aggressive beings that he portrayed. His "weapons" were found to be bones gnawed by hyenas, and the puncture wounds on the skulls were found to be the result of these early diminutive humans being carried away as prey by leopards.[4] The damage had already been done however. The image of the violent Indigenous past and ideas about the transformation of ape into a violent human was the basis for the popular movie *2001: A Space Odyssey.* It also influenced further ideas about what we might expect to find in the archaeological record. It wouldn't be the only time that we made false negative assumptions about distant hunter-gatherer ancestors.

An Archaeology of Our Fundamental Interdependence on Each Other

The idea that our ancestors *must* have been violent, competitive, and aggressive with one another to survive has been a pervasive one. However, evidence points instead to interdependence and collaboration with others, working together to find food, to defend themselves from predators, and to bring up astoundingly vulnerable and dependent young on which futures depended. Of course, people are complex, and there *is* evidence for violence in the past. For example, amongst Neanderthals living between 100,000 and 30,000 years ago, two individuals, one from Shanidar Cave in Iraq (with a projectile point injury), and one from St. Cesaire (with a frontal head injury), seem to have met a violent death at the hands of another person.[5] Even earlier, at around 450,000 BP, at Sima de los Huesos in Spain a man was also found to have most probably died from two blunt force head wounds, likely through interpersonal violence.[6] All people are bound to have been competitive at times, or even spiteful.[7] However, a record of strength through *interdependence on each other* in our distant past is far more extensive and compelling than that for aggression or competition.

Excavations across the world have revealed a past which speaks to us of a fundamental interdependence between people, built on truly human ways of reaching out and caring for others, which often contrasts sharply with our modern individualistic views of the world. One of my favorite examples of this mutual aid comes from a remarkable site at Man-Bac in Vietnam. Here, dating from around 4000 years ago, archaeologists found evidence of remarkable care for a disabled young man. This man was paraplegic, unable to move his arms or legs and barely able to move his head. His condition had started when he was juvenile, probably a decade before he died, yet for all this time he had been carefully cared for. The excavators concluded that he was completely dependent on those around him. Moreover, they concluded that looking after him must have involved a serious investment from his whole group—not just close family—in order to deal not only with his physical needs, but also with issues of his emotional wellbeing. Specific foods must have been prepared for him, and he must have been carefully turned regularly to avoid developing sores. Such care would be difficult to arrange even with modern medical facilities, yet clearly, as he was one of their own, his small-scale society would not consider abandoning him no matter how difficult his care was.[8]

Even earlier Indigenous hunter-gatherer communities also seem to have routinely looked after those who were ill or disabled. Several examples come from Europe during the last ice age, around 30,000 to 10,000 years ago. Few can doubt that such an environment was a difficult one in which to survive, and yet a man with dwarfism in Romito in Italy, dating from around 11,000 years ago, was clearly cared for by others despite the challenges.[9] Finding his own food is likely to have been difficult, not only due to his stature but to curvature of the spine and limited movement of his arms. Analysis demonstrated that he ate the same foods as others. During the ice age those with skeletal deformities were also carefully looked after and being cared for by the group was common both at the time of an injury and during a period of recovery.

Far earlier still, there is evidence that those who were ill and injured were looked after by their group. Neanderthals, our closest cousins in evolutionary time, also looked after their ill and injured, with most Neanderthals showing some injury and subsequent recovery.[10] For example, a Neanderthal man found at Shanidar cave in Iraq, dating from over 35,000 years ago, had a withered arm, a withered leg, and was probably blind in one eye. In short, he would have struggled to look after himself, but was nevertheless looked after by others, most probably the whole group, for well over a decade.[11] Likewise, individuals from Chapelle-aux-Saints and La Ferrassie in France, as well as a woman from Sale in Morocco with congenital deformations, were also carefully looked after, involving several people in their care.[12] Even as far back as 400,000 years ago in Sima de los Huesos in northern Spain where at least 29 people were deposited in a mortuary pit, evidence shows that an elderly man who could barely walk, a young child with a torsioned cranium, and a man who was deaf had clearly been cared for.[13] Evidence even exists for sustained care dating back over 1.5 million years ago in Africa where a woman found in Kenya suffered greatly from a terminal case of hypervitaminosis A and was clearly looked after for several weeks whilst in extreme pain.[14] Such accommodation and care for disabled people are rarely described in reconstructions of ancient peoples.

A *fundamental interdependence* between people is also seen in the remains of the far distant past in other ways. Emphasizing the interrelationship between people in methods of remembering, rather than distinctiveness or individuality, seems to have been important to early beliefs and ways of seeing the world. The earliest mortuary practices, for example, are not of "individuals" but involve deposition in certain ritual places where bones are mixed together, a practice which seems to reflect an ideology of collective identity. The earliest possible example of a mortuary practice comes from site Al-333 at Hadar in Ethiopia between 4 million and 2.5 million years ago. Here, the skeletal remains of at least 13 australopithecines were found together. Paul Pettitt argues that, rather than being brought together by any natural processes, they seem to have been deposited there because it was a recognized place to put the dead to rest.[15] Around 400,000 years ago, the approximately 28 individuals found at Sima de los Huesos in northern Spain were also deliberately interred together in a natural pit, this time with a possible "grave good," a rose colored hand axe.[16] A similar accumulation of early humans in a shared mortuary context around 300,000 to 200,000 years old has also been discovered in a cave at the Rising Star Cave system near Swartkrans in South Africa.[17] The practice of merging different people at death, almost as they might feel themselves part of each other in life, also extends right through the period of archaic humans and early human moderns and into later periods such as the Mesolithic and Neolithic. Modern Indigenous populations also often emphasize a continuity with others, with ancestors, and with nature at death.

There are other signs of ancestral attitudes and behaviors that stress *interdependence* rather than individuality. A lack of any indications that some people might have more material wealth than others until very late in the archaeological record has also been linked to an ancestral history of interdependent and egalitarian social systems such as that seen in many small-scale hunter-gatherers today. Egalitarianism and social relationships which *feel equal* have far reaching consequences, and are key to developing the type of playful interactions which allow children to flourish, and societies to be collaborative.[18]

Close emotional relationships and mutual trust seem to have been key to our survival and success as a species.[19] Of course life was by no means easy for our ancestors. Shortages of food were common, and injuries often resulted from hunting accidents or other types of dangers, even from predators.[20] Moreover, though rates of violence are typically low, both in archaeological contexts and in modern small scale hunter-gatherer groups, disputes doubtless occurred for which we find rare evidence.[21] One example comes from the ten skeletons dating from around 9,000 years ago found near Lake Turkana in East Africa, all of which showed signs of lethal aggression.[22] Nonetheless we now recognize that the fundamental interdependence of people that we see in the past and their willingness to help the vulnerable were key to surviving challenges together. Only in situations of mutual trust, where emotional commitments prompt people to willingly do things for others without considering the future reward, do people really build up relationships with the substantial give and take that it takes to weather hard times together.[23] It was interdependence, rather than individual self-interest, that allowed early humans to move into a new ecological niche through behaving, as Whiten and Erdal describe it, as "a single predatory organism."[24] Of course everyone has the capacity to be self-centered or altruistic, but whilst both ancient and modern Indigenous groups found ways to counter self-interest and promote pro-social tendencies at the expense of selfish ones, modern societies can often seem like a dangerous experiment in promoting individuality instead.

Certainly, what may be new to our societies, and rather alien to the distant past of Indigenous wisdom is *indifference*. For example, as Davi Kopenawa, a Yanomani shaman, travelled around New York, he was shocked by what he saw of people in need, abandoned by others. He said:

> Yet while the houses in the center of the city are tall and beautiful, those on its edges are in ruins. The people who live in those places have no food, and their clothes are

dirty and worn. When I took a walk among them, they looked at me with sad eyes. It made me feel upset. These white people who created merchandise think they are clever and brave. Yet they are greedy and do not take care of those among them who have nothing. How can they think they are great men and find themselves so smart? They do not want to know anything about these needy people, though they too are their fellows. They reject them and let them suffer alone. They do not even look at them and are satisfied to keep their distance and call them "the poor." They even take their crumbling houses from them. They force them to camp outside in the rain, with their children. They must tell themselves, "They live on our land, but they are other people. Let them stay far away from us, picking their food off the ground like dogs! As for us, we will pile up more food and more weapons, all by ourselves!" It scares me to see such a thing.[25]

For the vast majority of the human past, as in small scale societies today, people had little understanding or experience of *indifference* and were much more willing to give to those in need. Certainly, our ancient past hasn't prepared us for the inequalities of the modern world. Even the richest in any society have become unhappier as inequality rises.[26]

Clearly, while we often imagine a distant past of competitive individuals striving to survive against each other, evidence supports a different reality in which interdependence, and mutual support through vulnerabilities was the basis for human success. As much as suggesting that we should recognize this mutual personal interdependence, the archaeological record also supports a view of people in the distant past as having an interdependent—rather than a controlling—relationship with their environment.

An Archaeology of the Interdependence of People and Nature

In this volume, Narvaez et al. call the part of Indigenous wisdom that places us not as *better than*, but rather as *part of* the natural world, the "Philosophy of the Earth." This philosophy is much in evidence from earliest humans onwards. We often imagine ancient ancestors callously and thoughtlessly hunting prey. However, a relationship with animals that goes beyond the purely functional and suggests a certain respect dates back to at least 1.5 million years ago. For example, carefully made handaxes have been found on several sites across the world that were made from elephant bone, even though this material is difficult to work and functions less well than stone. Handaxes themselves are not only functional objects used for butchering meat, but are also made with great skill to a specific pleasing design that matches the "golden ratio" seen in architecture. They are more than simply practical tools. The fact that early humans chose to use elephant bone to make handaxes, even when stone was available, prompts Ran Barkai to argue that such handaxes reflect a deeper connection with elephants, perhaps based on the observation that these animals share with humans a capacity for empathy and complex social lives.[27] Other indications of a similar, respectful mentality occur in later archaic populations, as we see at sites in France and Italy from 90,000 years ago where Neanderthals used feathers from birds of prey in ways that seem highly symbolic.[28]

Some of the oldest realistic art also reflects a sense of intimate connection and even respect for nature. Realistic and sensitive depictions of animals which show a precise knowledge of their form and behavior date from around 40,000 years ago. Such art is particularly prolific in Ice Age Europe around 30,000 to 10,000 years ago. The artists must have very carefully observed the animals around them to create such images, knowing from memory exactly the shape of nostrils, patterns of markings, and stances of a wide variety of ice age animals (see Figure 2.1). Whatever else such art meant to the people who created it, it certainly meant that people were *still enough* to pay attention and understand the natural world around them in an intimate detail which we often fail to notice in our own busy lives.[29]

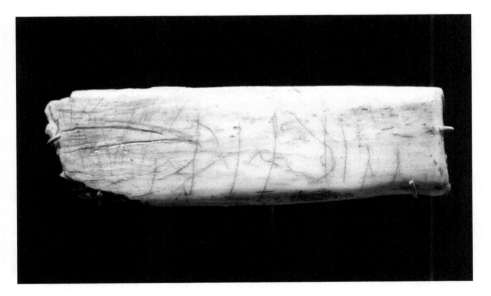

Figure 2.1. **Horse engraved on a bone from Creswell Crags**, around 12,500 years ago, and typically of the ubiquitous and sensitively created art from this period.
Source: Public Domain via Wikimedia Commons.

Cross cultural studies have even revealed a certain perceptual connection to nature shared by Indigenous societies that seems to have been lost in agricultural and industrialized contexts. Hunter-gatherers can visually perceive their surroundings in ways which those of us brought up in the "carpentered world" of modern buildings no longer can. For example, a common optical illusion (see Figure 2.2) experienced by people in modern western societies does not deceive the eyes of modern hunter-gatherers such as San bushmen.[30] Moreover, outside of the focus on *specifics* rather than *relationships* which modern schooling imposes people also see and understand the world in a more holistic way. Hunter-gatherer children with less school based learning and a greater emphasis on local ecological knowledge have been found to organize their understanding of the world around broader meaningful categories, whilst explicit schooling prompts a more sequential or rote learning organization of knowledge.[31] A holistic understanding of the world also leads to subtly different ways of drawing faces, with hunter-gatherers drawing faces in terms of feature blocks (such as with nose and eyebrows connected) rather than as individual spatially unconnected elements.[32] The environment we live in and how we learn profoundly affects how we see the world and how our minds organize and structure knowledge.[33]

Figure 2.2. **The Müller-Lyer illusion.** The lines labeled "a" and "b" are the same length. Many subjects perceive line "a" as longer than line "b" but hunter-gatherers such as San foragers are unaffected by the illusion.
Source: Author.

Like modern Indigenous peoples it seems that those in the distant past had a deep respect for nature. We are often told a story of an increasing "mastery" or control of nature in our human origins. However, we are only recently recognizing that our ancestral hunter-gatherer relationships with nature are better seen in terms of mutual understanding in functional as well as symbolic ways. New approaches to the relationship between people and the animals and plants in their environment can be seen in recent interpretations of the "domestication" of wolves. The gradual process by which wolves, from as early as 30,000 years ago, became integrated into human societies was once seen from the perspective of humans mastering the selection of traits in wolves which were functionally useful to them. However, we now realize that the process of wolves becoming "dogs" seems to have been as much due to the agency of wolves themselves as to human intentions, and perhaps as much due to emotional connections between the two species as wolves being used *for a function*. Wolves, themselves, are most likely to have chosen to shadow humans, exploiting possibilities to scavenge food with those who were least aggressive and most tolerant, eventually building up a relationship with people.[34] Wolves are increasingly recognized as part of a mutual interdependence with people rather than as being used as functional tools. For example, ethnographic studies have shown that, in many environments, dogs are not useful for hunting large game. Often they are used by women to hunt small game or are barely functional, mostly serving as sentries and living blankets.[35] Other evidence of an emotional interdependence between people and wolves/dogs include many examples of dog burials in ancient societies. In some Indigenous groups dogs are even seen as embodiments of ancestral people.[36] We know that modern dogs can act as attachment figures, able to create a sense of comfort and security in their owners, and these types of mutual bonds are likely to also have existed in the past. Rather than a mastering or controlling of the wild wolf, mutual respect and interdependence created today's constant companions.[37]

Even our long-held idea that humans and animals are fundamentally and distinctly different is increasingly seen as false; the evidence fits better with an Indigenous perspective that recognizes the fluidity of the boundaries between the two. When we look back on our origins as a species, there is no one place where animal becomes human, with different traits or features appearing at different times. For example, a modern human body form appears very early, but modern human brains are evident only at a later date. There is not even only one human, but rather many different species of human are present at any one time for almost all of our evolutionary past. Moreover, since species inevitably change through time it is clear that one day, whether by extinction or gradual change we will no longer exist as the species we are today. The lines between human and animal are as blurred in the reality of our evolutionary history as they are in traditional myths and stories.

Recognizing a deep past to our "philosophy of the earth" is important for our future. Whilst modern industrialized society portrays humans as having a dominion over nature which can be exploited for our own ends, recognizing that we are simply a part of nature changes that relationship. Acknowledging our vulnerability and dependence on the nature that we have left may change our views towards it. Even small reserves of nature, such as urban woodlands, have significant effects on stress, and decrease our tendencies to ruminate, or ponder issues in ways that are linked to depression and anxiety.[38] Now, as in the distant past, our relationship with nature is one of interdependence and mutual vulnerability.

A Sense of Awe

Lastly, it is easy to forget that experiencing a connection to the deep past of our ancestral hunter-gatherers can also be profoundly emotional. The clearest evidence of this is found in people's reactions to experiencing art from the distant past, which for many inspires awe at the continuity of human feeling over so many thousands of years. Experiencing our deep past can change our views of ourselves, making us feel more part of a wider society and more inclined to help others. Looking back on the evidence for humanity in the deep past also makes us awed by the sustainable wisdom that is the ideology which has supported us and our planet for 99% of our existence on earth.[39]

Despite how little survives of the lives of ancient hunter-gatherers who lived tens of thousands, hundreds of thousands, or even millions of years ago, what material evidence remains tells us about the fundamentals of our existence which resonate with Indigenous wisdom. Rather than competition, individuality, and control of nature, evidence suggests that a humble relationship with each other and with nature have made us successful as a species. Since the earliest humans arose, it seems that we have seen ourselves as part of others rather than self-oriented individuals, have been willing to care for those of us in the group who were vulnerable or in need, and have created egalitarian societies. Rather than mastering nature, we have been successful through respecting nature (as seen in certain types of tools, symbolic items and art), and in mutual relationships (as exemplified by how wolves moved into human groups). Both knowing about and experiencing what has been left to us from a deep past changes us, helping us to see our place in a continuity of human wisdom over thousands of generations.

Discussion Questions

How do you think that modern society influences assumptions about what life was like for hunter-gatherers in the distant past?

Why would hunter-gatherer communities of the distant past that cared for each other be more successful rather than ones where everyone was "in it for themselves"?

Why does our shared human heritage of distant hunting and gathering ancestors matter?

What could we do today to re-kindle a sense of connection and interdependance with each other and with nature?

Notes

1. Adam H. Boyette, "Children's Play and Culture Learning in an Egalitarian Foraging Society," *Child Development* 87, no. 3 (2016): 759–69.

2. For Inuit upbringing, see Jean L. Briggs, *Inuit Morality Play: The Emotional Education of a Three-Year-Old* (New Haven, CT: Yale University Press, 1999).

3. I vividly remember listening to descriptions of how ridicule is used in such societies to bring down those who seek to dominate others. Such egalitarian dynamics found in Mesolithic societies in Europe, as well as elsewhere, would later be one subject of my research, see Penny Spikins, "'The Bashful and the Boastful: Prestigious leaders and social change in Mesolithic Societies,'" *Journal of World Prehistory* 21 no. 3–4 (2008): 173–93; and Penny Spikins, "The Geography of Trust and Betrayal: Moral Disputes and Late Pleistocene Dispersal," *Open Quaternary* 1, no. 1 (2015): 1–10 http://www.openquaternary.com/articles/10.5334/oq.ai/print/.

4. See Penny Spikins, *How Compassion Made Us Human* (Barnsley, UK: Pen and Sword Books, 2015).

5. Christoph P. E. Zollikofer et al., "Evidence for Interpersonal Violence in the St. Césaire Neanderthal," *Proceedings of the National Academy of Sciences* 99, no. 9 (2002): 6444–48; Steven E. Churchill et al., "Shanidar 3 Neandertal Rib Puncture Wound and Paleolithic Weaponry," *Journal of Human Evolution* 57, no. 2 (2009): 163–78.

6. Nohemi Sala et al.,"Lethal Interpersonal Violence in the Middle Pleistocene," *PloS One* 10, no. 5 (2015): e0126589.

7. Penny Spikins, "The Geography of Trust and Betrayal."

8. For a description of Man Bac burial 9, see Marc F. Oxenham et al., "Paralysis and Severe Disability Requiring Intensive Care in Neolithic Asia," *Anthropological Science: Journal of the Anthropological Society of Nippon = Jinruigaku Zasshi* 117, no. 2 (2009): 107–12; and Lorna Tilley, "Care among the Neandertals: La Chapelle-Aux-Saints 1 and La Ferrassie 1 (Case Study 2)," in *Theory and Practice in the Bioarchaeology of Care* (New York, NY: Springer International Publishing, 2015), 191–218.

9. For descriptions of "the Romito dwarf," see W. Frayer David et al., "Dwarfism in an Adolescent from the Italian Late Upper Palaeolithic," *Nature* 330, no. 6143 (1987): 60; and Lorna Tilley, "Accommodating Difference in the Prehistoric Past: Revisiting the Case of Romito 2 from a Bioarchaeology of Care Perspective," *International Journal of Paleopathology* 8 (2015): 64–74. The stable isotope analysis revealing that this diet matched those of others is found in Oliver E. Craig et al., "Stable Isotope Analysis of Late Upper Palaeolithic Human and Faunal Remains from Grotta Del Romito (Cosenza), Italy," *Journal of Archaeological Science* 37, no. 10 (2010): 2504–12. For a description of the overall levels of injury and subsequent care see Xiu-Jie Wu et al., "Antemortem Trauma and Survival in the Late Middle Pleistocene Human Cranium from Maba, South China," *Proceedings of the National Academy of Sciences of the United States of America* 108, no. 49 (2011): 19558.

10. Hong Shang and Erik Trinkaus, "An Ectocranial Lesion on the Middle Pleistocene Human Cranium from Hulu Cave, Nanjing, China," *American Journal of Physical Anthropology* 135, no. 4 (2008):431–37.

11. For descriptions of the disabled man from Shanidar, see Erik Trinkaus and M. R. Zimmerman, "Trauma among the Shanidar Neandertals," *American Journal of Physical Anthropology* 57, no. 1 (1982): 61–76; and Erik Trinkaus, *The Shanidar Neandertals* (New York, NY: Academic Press, 1983).

12. For descriptions of Chapelle-aux-Saints and La Ferrassie, see Tilley, "Care among the Neandertals." For the woman from Sale in Morocco, see Jean-Jacques Hublin, "The Prehistory of Compassion," *Proceedings of the National Academy of Sciences of the United States of America* 106, no. 16 (2009): 6429–30.

13. For Sima de los Huesos see Alejandro Bonmatí et al., "Middle Pleistocene Lower Back and Pelvis from an Aged Human Individual from the Sima De Los Huesos Site, Spain," *Proceedings of the National Academy of Sciences of the United States of America* 107, no. 43 (2010): 18386; and Alejandro Bonmatí et al., "El Caso De Elvis El Viejo De La Sima De Los Huesos," *Dendra médica. Revista de humanidades* 10, no. 2 (2011): 138–46.

14. For the woman from 1.5 million years ago in Kenya called KNM-ER 1808, see A. Walker, M. R. Zimmerman, and R. E. F. Leakey, "A Possible Case of Hypervitaminosis A in Homo Erectus," *Nature* 296, no. 5854 (1982): 248.

15. For a discussion of AL-333, see Paul Pettitt, *The Palaeolithic Origins of Human Burial* (New York, NY: Routledge, 2011).

16. For Sima de los Huesos mortuary practices, see Eudald Carbonell et al., "Did the Earliest Mortuary Practices Take Place More Than 350,000 Years Ago at Atapuerca?" *Anthropologie* [Paris] 107, no. 1 (2003): 1–14.

17. For the Rising Star system remains, see Lee R. Berger et al., "Homo Naledi, a New Species of the Genus Homo from the Dinaledi Chamber, South Africa," *eLife* 4 (2015).

18. For a discussion of ancestral egalitarian social systems, see Christopher Boehm, *Moral Origins the Evolution of Virtue, Altruism, and Shame* (New York, NY: Basic Books, 2012); Christopher Boehm et al., "Egalitarian Behavior and Reverse Dominance Hierarchy" [and Comments and Reply], *Current Anthropology* 34, no. 3 (1993); Penny Spikins, "Goodwill Hunting? Debates over the Meaning of Lower Palaeolithic Handaxe Form Revisited," *World Archaeology* 44, no. 3 (2012): 378. For a discussion of how egalitarian influences playfulness and creativity, see Peter D. Gray, "The Play Theory of Hunter-Gatherer Egalitarianism," in *Ancestral Landscapes in Human Evolution: Culture, Childrearing and Social Wellbeing*, edited by Darcia Narváez, Kristin Valentino, Agustín Fuentes, James McKenna and Peter Gray. Oxford: Oxford University Press, 2014.

19. See Penny Spikins, Helen Rutherford, and Andy Needham, "From Hominity to Humanity: The Compassion from the Earliest Archaics to Modern Humans," *Time and Mind* 3, no. 3 (2010): 303–25; Spikins, *How Compassion Made Us Human*; Penny Spikins, "Prehistoric Origins: The Compassion of Far Distant Strangers," in *Compassion: Concepts, Research and Applications*, ed. Paul Gilbert (Didcot, UK: Taylor and Francis, 2017): 16–30; Brian Hare, "Survival of the Friendliest: Homo Sapiens Evolved Via Selection for Prosociality," *Annual Review of Psychology* 68 (2017): 155–86.

20. For evidence of injuries from foraging, see A. Lessa, "Daily Risks: A Biocultural Approach to Acute Trauma in Pre-Colonial Coastal Populations from Brazil," *International Journal Of Osteoarchaeology* 21, no. 2 (2011): 159–72.

21. Richard Wrangham, Michael Wilson, and Martin Muller, "Comparative Rates of Violence in Chimpanzees and Humans," *Primates* 47, no. 1 (2006): 14–26.

22. M. Mirazón Lahr et al., "Inter-Group Violence among Early Holocene Hunter-Gatherers of West Turkana, Kenya," *Nature* 529, no. 7586 (2016): 394.

23. See Randolph M. Nesse, *Natural Selection and the Capacity for Subjective Commitment*, (New York, NY: Russell Sage Foundation, 2001); and Spikins, "Goodwill Hunting?"

24. For a discussion of the origins of egalitarian collaboration, see Andrew Whiten and David Erdal, "The Human Socio-Cognitive Niche and Its Evolutionary Origins," *Philosophical Transactions of the Royal Society B: Biological Sciences* 367, no. 1599 (2012): 2119.

25. Davi Kopenawa, *The Falling Sky: Words of a Yanomami Shaman*, ed. Bruce Albert, Nicholas Elliott, and Alison Dundy (Cambridge, MA: The Belknap Press of Harvard University Press, 2013).

26. For varying levels of altruism in different societies, see Joseph Henrich, Steven J. Heine, and Ara Norenzayan, "Most People Are Not Weird," *Nature* 466, no. 7302 (2010): 29. For a discussion of how unequal societies make even the richest more unhappy, see Richard G. Wilkinson and Kate Pickett, *The Spirit Level: Why Equality Is Better for Everyone*, (London, UK: Penguin, 2009).

27. For a description, see Katia Zutovski and Ran Barkai, "The Use of Elephant Bones for Making Acheulian Handaxes: A Fresh Look at Old Bones," *Quaternary International* 406 (2016): 227–38; and Ma'Ayan Lev et al., "Elephants Are People, People Are Elephants: Human–Proboscideans Similarities as a Case for Cross Cultural Animal Humanization in Recent and Paleolithic Times," *Quaternary International* 406 (2016): 239–45.

28. For a description, see Eugène Morin and Véronique Laroulandie, "Presumed Symbolic Use of Diurnal Raptors by Neanderthals," *PLoS ONE* 7, no. 3 (2012): e32856.

29. For early art depictions in Sulawesi, see M. Aubert et al., "Pleistocene Cave Art from Sulawesi, Indonesia," *Nature* 514, no. 7521 (2014): 223. no. 7521 (2014 For overviews of upper palaeolithic art and the finding of cave art at Creswell Crags, see Paul G. Bahn, "Religion and Ritual in the Upper Palaeolithic," in *The Oxford Handbook of the Archaeology of Ritual and Religion*, ed. Tim Insoll (Oxford: Oxford University Press, 2011), 344–57; and Paul Bahn, Paul Pettitt, and Sergio Ripoll, "Discovery of Palaeolithic Cave Art in Britain," *Antiquity* 77, no. 296 (2003), 227–31.

30. For a description of how hunter-gatherers perceive their world, see Paul Rozin, "The Weirdest People in the World Are a Harbinger of the Future of the World," *Behavioral and Brain Sciences* 33, no. 2–3 (2010): 108–09. Rozin comments, in reference to American undergraduates, "On average, the undergraduates required that line "a" be about a fifth longer than line "b" before the two segments were perceived as equal. At the other end, the San foragers of the Kalahari were unaffected by the so-called illusion (it is not an illusion for them)."

31. Victoria Reyes-García et al., "Schooling, Local Knowledge and Working Memory: A Study among Three Contemporary Hunter- Gatherer Societies," *PLoS ONE* 11, no. 1 (2016): e0145265.

32. For a discussion of face drawing in hunter-gatherers and "western" populations, see Anneliese Pontius, "Spatial Representation, Modified by Ecology: From Hunter-Gatherers to City Dwellers in Indonesia," *Journal of Cross-Cultural Psychology* 24, no. 4 (1993): 399.

33. For a description of a more holistic view of the world shared by hunting and gathering populations, see Ayse K. Uskul, Shinobu Kitayama, and Richard E. Nisbett, "Ecocultural Basis of Cognition: Farmers and Fishermen Are More Holistic than Herders," *Proceedings of the National Academy of Sciences of the United States of America* 105, no. 25 (2008): 8552–56.

34. For a discussion of wolves building up a relationship with humans, see Brian Hare and Vanessa Woods, *The Genius of Dogs: How Dogs Are Smarter Than You Think* (New York, NY: Penguin, 2013).

35. For a discussion of the use of wolves by ethnographic populations, see Jane Balme and Susan O'Connor, "Dingoes and Aboriginal Social Organization in Holocene Australia," *Journal of Archaeological Science: Reports* 7 (2016): 775–81.

36. For a discussion of dog burials, see Darcy Morey, *Dogs Domestication and the Development of a Social Bond* (New York, NY: Cambridge University Press, 2010).

37. For a discussion of the use of wolves as attachment figures, see Lawrence A. Kurdek, "Pet Dogs as Attachment Figures," *Journal of Social and Personal Relationships* 25, no. 2 (2008): 247–66. and Miho Nagasawa et al., "Dog's Gaze at Its Owner Increases Owner's Urinary Oxytocin During Social Interaction," *Hormones and Behavior* 55, no. 3 (2009): 434–41.

38. For the effects of urban woodlands, see Liisa Tyrväinen et al., "The Influence of Urban Green Environments on Stress Relief Measures: A Field Experiment," *Journal of Environmental Psychology* 38 (2014): 1–9; and Gregory N. Bratman et al., "Nature Experience Reduces Rumination and Subgenual Prefrontal Cortex Activation," *Proceedings of the National Academy of Sciences of the United States of America* 112, no. 28 (2015): 8567.

39. See Michelle N. Shiota, Dacher Keltner, and Amanda Mossman, "The Nature of Awe: Elicitors, Appraisals, and Effects on Self-Concept," *Cognition & Emotion* 21, no. 5 (2007): 944–63.

References

Aubert, M., A. Brumm, M. Ramli, T. Sutikna, E. W. Saptomo, B. Hakim, M. J. Morwood, G. D. van den Bergh, L.Kinsley and A. Dosseto. "Pleistocene Cave Art from Sulawesi, Indonesia." *Nature* 514, no. 7521 (2014): 223–27.

Bahn, Paul G. "Religion and Ritual in the Upper Palaeolithic." In *The Oxford Handbook of the Archaeology and Anthropology of Hunter-Gatherers*, edited by Vicki Cummings, Peter Jordan, and Marek Zvelebil, 344–57. Oxford, UK: Oxford University Press, 2014.

Bahn, Paul, Paul Pettitt, and Sergio Ripoll. "Discovery of Palaeolithic Cave Art in Britain. (Research)." *Antiquity* 77, no. 296 (2003): 227–31.

Balme, Jane, and Susan O'Connor. "Dingoes and Aboriginal Social Organization in Holocene Australia." *Journal of Archaeological Science: Reports* 7 (2016): 775–81.

Berger, Lee R., John Hawks, Darryl J. de Ruiter, Steven E. Churchill, Peter Schmid, Lucas K. Delezene, Tracy L. Kivell, et al. "Homo Naledi, a New Species of the Genus Homo from the Dinaledi Chamber, South Africa." *eLife* 4 (2015).

Boehm, Christopher. *Moral Origins the Evolution of Virtue, Altruism, and Shame.* New York, NY: Basic Books, 2012.

Boehm, Christopher, Harold B. Barclay, Robert Knox Dentan, Marie-Claude Dupre, Jonathan D. Hill, Susan Kent, Bruce M. Knauft, Keith F. Otterbein, and Steve Rayner. "Egalitarian Behavior and Reverse Dominance Hierarchy." *Current Anthropology* 34, no. 3 (1993): 227–54.

Bonmatí, Alejandro, Asier Gómez Olivencia, Juan Luis Arsuaga, José Miguel Carretero Díaz, Ana Gracia, Ignacio Martínez, and Carlos Lorenzo Merino. "El Caso De Elvis El Viejo De La Sima De Los Huesos." *Dendra médica. Revista de humanidades* 10, no. 2 (2011): 138–46.

Bonmatí, Alejandro, Asier Gómez-Olivencia, Juan-Luis Arsuaga, José Miguel Carretero, Ana Gracia, Ignacio Martínez, Carlos Lorenzo, José María Bérmudez de Castro, and Eudald Carbonell. "Middle Pleistocene Lower Back and Pelvis from an Aged Human Individual from the Sima De Los Huesos Site, Spain." *Proceedings of the National Academy of Sciences of the United States of America* 107, no. 43 (2010): 18386–91.

Boyette, Adam H. "Children's Play and Culture Learning in an Egalitarian Foraging Society." *Child Development* 87, no. 3 (2016): 759–69.

Bratman, Gregory N., J. Paul Hamilton, Kevin S. Hahn, Gretchen C. Daily, and James J. Gross. "Nature Experience Reduces Rumination and Subgenual Prefrontal Cortex Activation." *Proceedings of the National Academy of Sciences of the United States of America* 112, no. 28 (2015): 8567–72.

Briggs, Jean L. *Inuit Morality Play: The Emotional Education of a Three-year-old.* New Haven, CT: Yale University Press, 1999.

Carbonell, Eudald, Marina Mosquera, Andreu Ollé, Xosé Rodríguez, Robert Sala, Josep Vergès, Juan Arsuaga, and José-María Bermúdez de Castro. "Did the Earliest Mortuary Practices Take Place More Than 350,000 Years Ago at Atapuerca?" *Anthropologie* [Paris] 107, no. 1 (2003): 1–14.

Churchill, Steven E., Robert G. Franciscus, Hilary Mckean-Peraza, Julie A. Daniel, and Brittany R. Warren. "Shanidar 3 Neandertal Rib Puncture Wound and Paleolithic Weaponry." *Journal of Human Evolution* 57, no. 2 (2009): 163–78.

Craig, Oliver E., Marco Biazzo, André C. Colonese, Zelia Di Giuseppe, Cristina Martinez-Labarga, Domenico Lo Vetro, Roberta Lelli, Fabio Martini, and Olga Rickards. "Stable Isotope Analysis of Late Upper Palaeolithic Human and Faunal Remains from Grotta Del Romito (Cosenza), Italy." *Journal of Archaeological Science* 37, no. 10 (2010): 2504–12.

David, W. Frayer, A. Horton William, Macchiarelli Roberto, and Mussi Margherita. "Dwarfism in an Adolescent from the Italian Late Upper Palaeolithic." *Nature* 330, no. 6143 (1987): 60–62.

Gray, Peter D. "The Play Theory of Hunter-Gatherer Egalitarianism." In *Ancestral Landscapes in Human Evolution: Culture, Childrearing and Social Wellbeing*, edited by Darcia Narváez, Kristin Valentino, Agustín Fuentes, James McKenna and Peter Gray, 192–215. Oxford: Oxford University Press, 2014.

Hare, Brian. "Survival of the Friendliest: Homo Sapiens Evolved Via Selection for Prosociality." *Annual Review of Psychology* 68 (2017): 155–86.

———, and Vanessa Woods. *The Genius of Dogs: How Dogs Are Smarter Than You Think.* New York, NY: Penguin, 2013.

Henrich, Joseph, Steven J. Heine and Ara Norenzayan. "Most People Are Not Weird." *Nature* 466, no. 7302 (2010): 29.

Hublin, Jean-Jacques. "The Prehistory of Compassion." *Proceedings of the National Academy of Sciences of the United States of America* 106, no. 16 (2009): 6429–30.

Kopenawa, Davi. *The Falling Sky: Words of a Yanomami Shaman.* Edited by Bruce Albert, Nicholas Elliott, and Alison Dundy. Cambridge, MA: The Belknap Press of Harvard University Press, 2013.

Kurdek, Lawrence A. "Pet Dogs as Attachment Figures." *Journal of Social and Personal Relationships* 25, no. 2 (2008): 247–66.

Lessa, A. "Daily Risks: A Biocultural Approach to Acute Trauma in Pre-Colonial Coastal Populations from Brazil." *International Journal of Osteoarchaeology* 21, no. 2 (2011): 159–72.

Lev, Ma'Ayan, and Ran Barkai. "Elephants Are People, People Are Elephants: Human–Proboscideans Similarities as a Case for Cross Cultural Animal Humanization in Recent and Paleolithic Times." *Quaternary International* 406 (2016): 239–45.

Mirazón Lahr, M., F. Rivera, R. K. Power, A. Mounier, B. Copsey, F. Crivellaro, J. E. Edung, et al. "Inter-Group Violence among Early Holocene Hunter-Gatherers of West Turkana, Kenya." *Nature* 529, no. 7586 (2016): 394–98.

Morey, Darcy. *Dogs: Domestication and the Development of a Social Bond.* New York, NY: Cambridge University Press, 2010.

Morin, Eugène, and Véronique Laroulandie. "Presumed Symbolic Use of Diurnal Raptors by Neanderthals." *PLoS ONE* 7, no. 3 (2012): e32856. https://doi.org/10.1371/journal.pone.0032856.

Nagasawa, Miho, Takefumi Kikusui, Tatsushi Onaka, and Mitsuaki Ohta. "Dog's Gaze at Its Owner Increases Owner's Urinary Oxytocin During Social Interaction." *Hormones and Behavior* 55, no. 3 (2009): 434–41.

Nesse, Randolph M. *Natural Selection and the Capacity for Subjective Commitment.* New York, NY: Russell Sage Foundation, 2001.

Oxenham, Marc F., Lorna Tilley, Hirofumi Matsumura, Lan Cuong Nguyen, Kim Thuy Nguyen, Kim Dung Nguyen, Kate Domett, and Damien Huffer. "Paralysis and Severe Disability Requiring Intensive Care in Neolithic Asia." *Anthropological Science: Journal of the Anthropological Society of Nippon = Jinruigaku Zasshi* 117, no. 2 (2009): 107–12.

Pettitt, Paul. *The Palaeolithic Origins of Human Burial.* New York, NY: Routledge, 2011.

Pontius, Anneliese. "Spatial Representation, Modified by Ecology: From Hunter-Gatherers to City Dwellers in Indonesia." *Journal of Cross-Cultural Psychology* 24, no. 4 (1993): 399–419.

Reyes-García, Victoria, Aili Pyhälä, Isabel Díaz-Reviriego, Romain Duda, Álvaro Fernández-Llamazares, Sandrine Gallois, Maximilien Guèze, and Lucentezza Napitupulu. "Schooling, Local Knowledge and Working Memory: A Study Among Three Contemporary Hunter-Gatherer Societies." *PloS one* 11, no. 1 (2016): e0145265.

Rozin, Paul. "The Weirdest People in the World Are a Harbinger of the Future of the World." *Behavioral and Brain Sciences* 33, no. 2–3 (2010): 108–09.

Sala, Nohemi, Juan Luis Arsuaga, Ana Pantoja-Pérez, Adrián Pablos, Ignacio Martínez, Rolf M. Quam, Asier Gómez-Olivencia, José María Bermúdez de Castro, and Eudald Carbonell. "Lethal Interpersonal Violence in the Middle Pleistocene." *PloS One* 10, no. 5 (2015): e0126589.

Shang, Hong, and Erik Trinkaus. "An Ectocranial Lesion on the Middle Pleistocene Human Cranium from Hulu Cave, Nanjing, China." *American Journal of Physical Anthropology* 135, no. 4 (2008): 431–37.

Shiota, Michelle N., Dacher Keltner, and Amanda Mossman. "The Nature of Awe: Elicitors, Appraisals, and Effects on Self-Concept." *Cognition & Emotion* 21, no. 5 (2007): 944–63.

Spikins, Penny. "'The Bashful and the Boastful': Prestigious leaders and social change in Mesolithic Societies." *Journal of World Prehistory* 21, no. 3–4 (2008): 173–93.

———. "The Geography of Trust and Betrayal: Moral Disputes and Late Pleistocene Dispersal." *Open Quaternary* 1, no. 1 (2015): 1–10. http://www.openquaternary.com/articles/10.5334/oq.ai/print/.

———. "Goodwill Hunting? Debates over the Meaning of Lower Palaeolithic Handaxe Form Revisited." *World Archaeology* 44, no. 3 (2012): 378.

———. *How Compassion Made Us Human.* Barnsley, UK: Pen and Sword Books, 2015.

———. "Prehistoric Origins: The Compassion of Far Distant Strangers." In *Compassion: Concepts, Research and Applications*, edited by Paul Gilbert, 16–30. Didcot, UK: Taylor and Francis, 2017.

Spikins, Penny, Helen Rutherford, and Andy Needham. "From Hominity to Humanity: Compassion from the Earliest Archaics to Modern Humans." *Time and Mind* 3, no. 3 (2010): 303–25.

Tilley, Lorna. "Accommodating Difference in the Prehistoric Past: Revisiting the Case of Romito 2 from a Bioarchaeology of Care Perspective." *International Journal of Paleopathology* 8 (2015): 64–74.

———. "Care among the Neandertals: La Chapelle-Aux-Saints 1 and La Ferrassie 1 (Case Study 2). " In *Theory and Practice in the Bioarchaeology of Care*, 219–57. New York, NY: Springer International Publishing, 2015.

Trinkaus, Erik. *The Shanidar Neandertals.* New York, NY: Academic Press, 1983.

Trinkaus, Erik, and M. R. Zimmerman. "Trauma among the Shanidar Neandertals." *American Journal of Physical Anthropology* 57, no. 1 (1982): 61–76.

Tyrväinen, Liisa, Ann Ojala, Kalevi Korpela, Timo Lanki, Yuko Tsunetsugu, and Takahide Kagawa. "The Influence of Urban Green Environments on Stress Relief Measures: A Field Experiment." *Journal of Environmental Psychology* 38 (2014): 1–9.

Uskul, Ayse K., Shinobu Kitayama, and Richard E. Nisbett. "Ecocultural Basis of Cognition: Farmers and Fishermen Are More Holistic than Herders." *Proceedings of the National Academy of Sciences of the United States of America* 105, no. 25 (2008): 8552–56.

Walker, Alan, Michael R. Zimmerman, and Richard E. F. Leakey. "A Possible Case of Hypervitaminosis a in Homo Erectus." *Nature* 296, no. 5854 (1982): 248–50.

Whiten, Andrew, and David Erdal. "The Human Socio-Cognitive Niche and Its Evolutionary Origins." *Philosophical Transactions of the Royal Society B: Biological Sciences* 367, no. 1599 (2012): 2119–29.

Wilkinson, Richard G., and Kate Pickett. *The Spirit Level: Why Equality Is Better for Everyone.* London, UK: Penguin, 2009.

Wrangham, Richard, Michael Wilson, and Martin Muller. "Comparative Rates of Violence in Chimpanzees and Humans." *Primates* 47, no. 1 (2006): 14–26.

Wu, Xiu-Jie, Lynne A. Schepartz, Wu Liu, and Erik Trinkaus. "Antemortem Trauma and Survival in the Late Middle Pleistocene Human Cranium from Maba, South China." *Proceedings of the National Academy of Sciences of the United States of America* 108, no. 49 (2011): 19558–62.

Zollikofer, Christoph P. E., Marcia S. Ponce de León, Bernard Vandermeersch, and François Lévêque. "Evidence for Interpersonal Violence in the St. Cesaire Neanderthal." *Proceedings of the National Academy of Sciences of the United States* 99, no. 9 (2002): 6444–48.

Zutovski, Katia, and Ran Barkai. "The Use of Elephant Bones for Making Acheulian Handaxes: A Fresh Look at Old Bones." *Quaternary International* 406 (2016): 227–38.

CHAPTER THREE

Indigenous Bodies, Civilized Selves, and the Escape from the Earth

EUGENE HALTON

Prelude: Walking Wisdom

About a half century ago a friend of mine who later became an anthropologist, Thomas "Tim" Buckley (1942–2015), was apprenticed to and adopted as a nephew by Harry Roberts (1906–81), a Yurok spiritual teacher, in northern California. Among the many things he learned, as Buckley put it, Roberts

> taught us that when we walk, we can feel the earth pushing back, supporting us. We talk about loving the earth. But we can also feel the earth loving us. Pay attention. Pay attention to the stability and the reliability of the earth. You can say "love." Harry would never use that word. But I use it.[1]

Buckley learned a whole different awareness of our place on earth. I now feel these words from my friend's uncle when walking, I feel how one's energy is immediately participant in one's surroundings—including those that extend beneath us into the earth—through step-wise resistance, as well: walking within the earth rather than upon it.

Recently I meditated on a question of how humanity can make a difference in the downward trends of the biosphere. I soon visualized Atlas holding the entire world on his shoulders (in the ancient myth he held the celestial spheres, but

I was visualizing the modern misconception that he held the earth up). As my vision continued, mighty Atlas put the earthly globe down from his shoulders, and lay on it on his back (as with a kind of exercise ball), stretching. The image was humorous, but I allowed the process to continue. Atlas began walking on the globe, leaving an enormous "foot print," and I saw our human relation to the earth as that seemingly crushing domination.

Still in my vision, I realized that the human needed to be put into perspective for the good of the earth, and so Atlas began to get smaller as the earth got larger. As the process progressed I became Atlas walking the earth, increasingly grasping the smallness of the human in relation to the great globe, progressively feeling the sustaining power of the earth upon my walking feet in each step. Things clarified: The earth holds us; we, in our "titanic" myths of ego and machine power, do not hold it. The extrapolation of the human to titanic proportions—to Atlas holding the world or the celestial spheres, as though they depend on his power—is a delusion that only speaks to our arrogant attempts to escape the laws of creation. Through our fascination with the myth of the machine, we ignore how those laws also include "the earth pushing back, supporting us ... the earth loving us."

From the Philosophy of the Earth to Escape from the Earth: Up, Up, and Away!

I will show throughout this chapter that history needs to be understood as involving a problematic interplay between the long-term legacy of human evolution, still tempered into the human body today, and the shorter-term heritage of civilization from its beginnings to the present. Each of us lives in a tension between our Indigenous bodies and our civilized selves, between *the philosophy of the earth* and that which I characterize as *"the philosophy of escape from the earth."* The standard story of civilization is one of linear upward progress, a story that I will contest with an alternative philosophy of history that I have developed, picturing history instead as a set of concentric circles.

One simple way to view the gap between first ways foragers and civilized and modern ways is through the opposed worldviews of *the philosophy of the earth* versus *the philosophy of escape from the earth.* Humans evolved into being as foragers under the guidance of *the philosophy of the earth.* We evolved into humanity through the close attunement to circumambient life, an outlook folklorist and philosopher John Stuart-Glennie characterized as *panzoonism,* belief in the living powers of all things. In this sense we are children of the earth, forged from the combined practical and reverential attunement to it, as I put it in coining the term *sustainable wisdom* in 2013:

> Though we may think ourselves modern, we retain Pleistocene bodies ... and Pleistocene needs, bodied into being over our longer two million year evolution. What Shepard termed "the sacred game," the dramatic interplay of predator and prey, reminds us of that older evolutionary story, wherein [humanity] emerges into being wide-eyed in wonder at circumambient life, a child of the earth foraging for edible, sensible, thinkable, and sustainable wisdom.[2]

Through the domestication of plants and animals, and through permanent settlements, the rise of agriculture and civilization marked a radical transition toward a new outlook, one built on an idea of controlling nature toward human ends, but which I term *the philosophy of escape from the earth*.

Let us consider a few of the countless variations of the linear progress extrapolation graph: a line from the bottom left, ever so slightly beginning to lift, then the upswing of the slope, and finally "up, up, and away"—things get progressively better. One version shows not much happening for the last 2 million years, a slight lift in the past 50,000 years, an upswing in the past 10,000 years with agriculture and then cities, and then "up, up, and away" with the Industrial Revolution. Thanks to Western civilization in the modern era, as the story goes, things begin to go up exponentially and continue to elevate globally. Those peoples colonized, enslaved or exterminated by imperializing, industrial civilization would of course view this story quite differently.

A variant graph I came across traced the evolution of the most significant inventions, with the Bronze Age to the Iron Age showing a slight lift, and the trajectory escalating in post-medieval Europe. Between 1281 and 1600 only gunpowder and magnetism are listed. Unfortunately this graph left out the mechanical clock, which Lewis Mumford called "the key machine of the modern industrial age." And that is not to mention the novel, which Czech novelist Milan Kundera claimed as one of the most important inventions of the modern era, opening up a new inner landscape. Nevertheless the trajectory through plastics, airplanes, radar, computers, and atomic energy still resulted in a happy "up, up and away" exponential ending—literally—with its last listed item: space travel.[3]

Computational power versus human brain power represents another variant of the civilization-as-ascending-progress story. One can imagine this graph, imbued with the utopian visions of the futurists, artificial intelligence advocates, and other technophiles, having the same trajectory. By about 2023 the human brain is supposed to be surpassed by computers. And to those who might worry about the possibility of gradual absorption of one's mind by the devices that take one and one's children completely into their screens: not to worry. By 2045, according to futurist Ray Kurzweil, the machine will be doing better than all of humanity combined. Paradise, the Promised Land, will bring deliverance as *the singularity*: "The Singularity will allow us to transcend these limitations of our biological bodies and brains– ... There will be no distinction, post-Singularity, between human and machine."[4] I can think of some distinctions Kurzweil neglects: the machine will likely not be put out of work by automation, poor people will not be absorbed into singularity elites, and world devouring capitalism is not likely to divert sufficient profits or develop a "seven generations capitalism" model to heal the biosphere.

If living were simply computation, we could even go further and completely "upload" ourselves from this world, to total electronic virtuality, and finally be done with the earth. But living is so much more than the schizoid model of the machine. Notice how these pro-progress scenarios make no mention of the conditions required to support and sustain complex organic life. Such views ignore how the earth itself is generative, not reducible to the model of the machine. Earthiness is simply assumed as something to be transcended through power and technical control; as though human embodiment is inferior to disembodied cyberspace, and quantified electronic information is superior to the palpable poetic wonder of the variescent earth. Philosopher and physicist Charles Peirce coined the term *variescence* to claim that the universe is not governed completely by immutable law, but that there is a real element of spontaneity through which new variety can come into being. It implied that laws of nature evolve, and that evolution itself can evolve.[5]

Here is a final variant from historian Henry Adams' 1909 essay, "The Rule of Phase Applied to History" (Figure 3.1), with the same slope going "up, up and away." Adams was trying to map the increase in power over time, and he saw how the invention of the steam engine around 1800 radically increased power, and precision as an instrument of power.

Though the slope of his graph is similar to the others, Adams realized what none of the people in his age, and virtually none of the technophile, technomaniacs of our age consider: that the increase in power is more than simply machines or technology, it includes the context of the social institutions surrounding it. Thus as power in machines increases, corresponding increases in the controls provided by social institutions are required. Adams intuited what Lewis Mumford later spelled out: the idea that power can be reduced to technological functions independent of human culture is an illusion that misses how technology is a human production requiring human purposes in mind, rather than an autonomous realm with rules of its own that we simply have to come into accord with.[6]

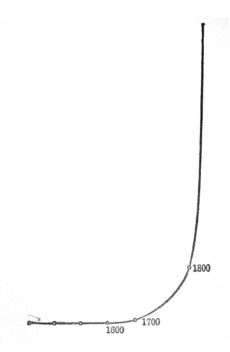

Figure 3.1: Henry Adams, the increase in power over time.
Source: H. Adams, "The Rule of Phase Applied to History." in H. Adams & B. Adams, *The Degradation of the Democratic Dogma* (New York, NY: Macmillan, 1919), 267–311.

Along with the above described trajectory upward slope comes population explosion, with its devastations as well, including climate change. Consider that our most intimate relationship with the environment is through daily food. Civilization is based on a fundamental alteration of food, the "agricultural revolution," and its population exploding consequences have never ended. The so-called "Green revolution,"—both the use of phosphate fertilizers and the introduction of high yield wheat developed by Norman Borlaug—from about 1948 on, amped up the agricultural revolution and its population explosion into a much greater extrapolation upward, especially in India and Pakistan. Borlaug was credited with "saving lives," but the radical increases in population (in India and Pakistan especially) and the resultant threats of malnutrition and resource depletion were also direct consequences.

Between 1970 and 2018 human population doubled, from a little over 3.5 billion to over 7.7 billion people. In roughly the same time period, between 1970 and 2014, there was a global average decline of 60 percent of vertebrate species populations, and regionally an 89 percent decline in Central and South America. Almost 44% of children in Pakistan today are stunted, physically and mentally, due to malnutrition. That is over 28 million children. In 2015–2016 a third global coral bleaching event occurred (the first two were in 1998 and 2010), due to very warm El Niño ocean temperatures, and was the longest, most widespread, and most damaging thus far, with severe threats to the Australian Great Barrier Reef as well as many others. Numerous scenarios of unsustainability puncture the happy "up, up, and away" ascent of the myth of progress, and raise the question of the role of science and technology as manifestations, not of progress, but of a ruinously self-destructive logic.

Modern science and technology emerged as a vast social construction, an outlook that the universe is an enormous machine, a grand clockwork, embodying what Lewis Mumford characterized in his two-volume book as the "myth of the machine."[7] This false myth—that the machine is an autonomous force independent of human purpose—has not prevented science from discovering countless arrays of precise natural facts and natural constructions, and of transforming the world, but in its classic form it stunts the reality of science, of nature and sociality. It undervalues or excludes elements of reality that do not fit the reductionist machine-like characteristics.[8]

Despite the vast achievements and transformations of the earth brought about by science and technology, Charles Peirce claimed that the "scientific realist" framework in which science emerged harbors illogical assumptions ultimately at odds with realism and with the continued development of science. As a logician and practicing scientist Peirce challenged science to come to terms with a more comprehensive living universe, alive in still active creation and a reasonableness energizing into being.[9]

The upsurge of quantifying precision (in money and its measures and in mechanized manufacture and science) symbolized by the mechanical clock, led to a skewed worldview where quantity displaced the qualitative aspects of experience, where the model of the mechanical machine displaced the model of organic life. These changes reflected the transformation of Western and now global consciousness toward one that I characterize as *mechanico-centric*, as I will describe later.

Nevertheless, it must be admitted that genuine progress in the sciences and technology not only exists, but that the practices of science and technology are absolutely requisite for addressing the very problems of unsustainability to which they may have contributed. The emerging scientific concept of the Anthropocene, for example, draws the long record of climate and life into awareness to reveal the significant human-influenced impacts on global ecosystems, and impact now on a potential sixth major period of extinctions.

So, how can we reconcile the facts of true progress in modern civilization with the genuine increases in unsustainability? How can we reconcile sustainable wisdom with unsustainable folly? Ecological philosopher Paul Shepard suggested the direction: "The tools we have invented for communicating our ideas and carrying information have actually impaired our memories. We must begin by remembering beyond history."[10] Remembering beyond history involves remembering the evolutionary trajectory that created the human body—not simply as some proto-utilitarian calorie counter, as too many biological accounts would have it, but as a creature alive to the moment practically, aesthetically, spiritually, and socially.

The intelligible wild habitat, engaged through a ritual participatory attitude that Shepard has characterized as "the sacred game," nurtured the evolution of humans through attunement given to it. By "the sacred game," Shepard meant the primal importance of the relationship between predator and prey, the wild other and the human, in actual hunting and gathering practices. And it retains a central significance as well in the varied activities of life, including ritual, parenting, individual and clan identity, and as a mode of imagination and thought, wherein the wild other is a focal source of attunement.

Ethnographic accounts provide substantial evidence that wild animals and plants held a central place in the practical and imaginative lives of human foragers. The accounts also depict a "participation attitude," a consciousness of being passionately attuned to the ecological intelligence of the community of life.[11] This participation attitude, synonymous with what has been called animism (perhaps more aptly called panzoonism) involves a relational consciousness, thoroughly involved in its living and signifying habitat. In this relational outlook, things are not inanimate substances, but rather animate signs through which one finds clues and cues for living. The bodies we evolved into, including psyche and self, were nourished by the omnivorous engagement with the wild other in the sacred game.

It makes great sense to live in an ongoing dialogue with the intelligible signs of the habitat, and to respect the potential wisdom it can impart. As biologist Robin Wall Kimmerer puts it,

> To be native to a place we must learn to speak its language ... Learning the grammar of animacy could well be a restraint on our mindless exploitation of land ... [it can] remind us of the capacity of others as our teachers, as holders of knowledge, as guides. Imagine walking through a richly inhabited world of birch people, bear people, rock people, beings we think of and therefore speak of as persons worthy of our respect, of inclusion in a peopled world. ... Imagine the access we would have to different perspectives, the things we might see through other eyes, the wisdom that surrounds us.[12]

In other work, I have argued that the increased, prolonged, unmatured characteristics of neoteny that distinguish humans from other primates may require attunement to wild, mature nature in order for humans to achieve their own maturity.[13] The "native state" of the plastic human mind as it evolved into being involves the rationality derived from the animal and plant minds to which it is omnivorously attuned, and through which it can find its own maturity. Rationality is not simply a product of human consciousness, but, as Peirce claimed, it is that "to which experience and reflexion would tend indefinitely to make human approval conform."[14] Animal minds can be focused, attentive, and experiential, and plants and fungi reveal health promoting, symbiotic social networks, a "quasi-mind," if you will.[15]

If humans are supposed to be more reflective creatures, that reflectivity was originally likely expressive reflectivity, ritually based, something like entrained, affirmative thought. Dancing an animal or tracking an animal, for example, puts you in its habits of mind, and puts those habits of mind into human social forms. Tracking involves literally a minute logic, a complex syntax of a micro-landscape involved in any given track, as well as a capacity for entrainment: it is a practical science and art.

Human plasticity and neoteny, together with progressively cooperative social behaviors, required increased intersociality involving the social mind of the living habitat as its food for thought, literally and metaphorically. This capacity to find maturity through attunement to "the generalized wild other" (to paraphrase George Herbert Mead's term "the generalized other") was diminished in the transformation effected by domestication, settlement, and civilization. Its last outpost, so to speak, was nascent science, which, unfortunately, has remained too long under the spell of the machine, the "generalized mechanical other." Yet Peirce's view of rationality and mind opens the scientific door to the rationality to be found in wild nature, not simply as the object of scientific investigation, but as profoundly formative of the mind of science.

With the advent of agriculture and civilization, human societies turned from the wild other as a central role model to other humans, and, in the modern era, they have turned progressively to machine mediated interaction and machine models of mind. When humans start to treat other neotenous, unmatured primates, that is, other humans, as ultimate role models, the loss of wild attunement represents a tragic escape from the informing philosophy of the earth that nurtured us into being and that remains tempered into the human body, psyche, and developmental needs.

In short, the rise of agriculturally based civilizations globally introduced a radical departure from the conditions that sustained the emergence of anatomically modern humans. The civilized self contracted from the centrality of direct engagement with wild nature to a progressively domesticated, human dominant outlook. From its very beginnings, the civilized self embarked on a new direction, that of the philosophy of escape from the earth, a journey that continues unabated and even amplified to the present day. Framing this journey as progress apparently still convinces most people that escaping the earth is a worthy way of living.

The Contractions of Mind

The Primeval State of Man, was Wisdom, Art, and Science.

—William Blake[16]

Most philosophies of history since the Enlightenment assume progress. Hegel's dialectical model takes an idea as eventually giving rise to a counter-position, and a third "synthesis" as providing a resolution of those contrasts and a new, more comprehensive phase. Comte also took a three-part view of history, understanding it as a progression from an original theological or fictitious state, to a more abstract outlook of a metaphysical state, to a third and final scientific or positive state. I have devised a new philosophy of history with a different three-part approach to understanding human development, taking civilization not as a linear advance of progress, but rather as a progress in precision, paradoxically counteracted by a regression in mind: history as a contraction of mind.[17]

I describe three stages in the contraction of mind: 1) *animate mind* as the evolved outlook of foraging life; 2) *anthropocentric mind* as representing the contracting transformation of consciousness produced by agriculturally-based civilization; 3) *mechanico-centric mind* as representing a further contraction from human-centered to a machine-centered consciousness, produced by the rise of modern civilization and the mechanical scientific worldview. Hence, this progressive contraction is marked by a turn from original practical and reverential attunement to the living earth in hunting and gathering societies, or *animate mind*, to a narrower focus of *anthropocentric mind* beginning with the development of early civilizations, where the human element became central and the wild devalued. And it moves to an even more narrow focus of *mechanico-centric mind*, expanding out of late medieval Europe and the development of modern science, where the machine became model of the ultimate, the objectivist filter through which the world is to be understood and made to fit.

The two million years of foraging through which we evolved were the bodying forth and shaping of what I term animate mind. As Figure 3.2 illustrates, animate mind is that evolutionary mindset reverentially and practically attuned to circum-ambient life, attuned to the *animate earth*.[18]

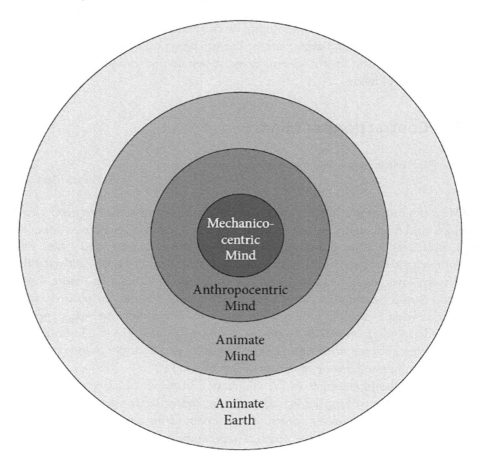

Figure 3.2: The Contractions of Mind.
Source: Author.

The development of agriculturally-based civilization marked a contraction, a progressive contraction from our relation to the living earth (the life and habitat around us and our relation to the wild significant other as key in human outlook and beliefs) to the human as the primary significant other, and, eventually to the human elevated as a divine other. In that process, we walled ourselves off from the wild with agriculture and civilization. We domesticated animals and plants to human purposes, that is, to anthropocentric purposes. We achieved progress in precision, in farming and animal husbandry, in city building and architecture, in bureaucratic organization of military, laboring, and scribal institutions, even as direct attunement to the animate earth receded.

Long before the modern clockwork universe, in the thousands of years in which cities first came into being, new ways of consciousness emerged in a transformation of humans into city-living, civilized beings. The city center became the virtual, and indeed, the sacred pivot of the universe in many cultures, even as the wild habitat lost significance.

The advent of civilization brought with it the rise of the *spectator attitude*, an attitude in which the human spectacle took center stage. Even well before the first cities, about 10,000 BCE, hunter gatherers on the verge of domestication and agriculture at Göbekli Tepe in Southern Turkey created the first known monumental complex, centered around Megalithic stones shaped in a human T form, 4–5 meters (13–16 feet) tall, and weighing between 5 to 10 tons. Some have arms and fingers. Animals are also incised on them, including boars, foxes, serpents, etc. As archaeologist Ian Hodder noted:

> There are lions, deer, and cattle, and also scorpions and all sorts of things. These are on the human (forms). And, so the human has become central. And I think that is the most important shift that we see. Because it's only when humans become central to the natural world and become able to dominate animals that they can domesticate animals.[19]

As excavator Klaus Schmidt put it, "First came the temple, then the city." Humans created a ritual self-reflection in the first monumental temples, though still heavily adorned with the imprint of the animal.[20]

With the rise of cities thousands of years later, the temple and palace become the *axis mundi*, embodying the city as the pivot of the world, and the king as its semi-divine or divine apex. The "celestial spheres" of the night sky become the personified spectacle of the Greek Titan Atlas, condemned by Olympian Zeus to hold them aloft. Monotheism later projects a transcendent divinity "out of this world."

The course of civilization was to "elevate" engagement toward spectatorship, and away from participation; toward a drama of a transcendent, supernatural realm beyond nature, and away from the living and immanent cosmos; toward "mind" and its conceptions of gods, and away from the direct touch of the earth. The mind of civilization puts the human drama into the foreground, early on this is done through conceptions of gods and divine kings, prophets, saviors, and enlightenment, and later through myths of materialism and machines as ultimate, as providing the fundamental lessons to be learned.

Basic to civilization since its beginnings in the domestication of animals and plants has been the attempt to control nature. The technologies of agriculture sought to wall in the wild through domesticated confinement, in order to provide the populace with sustainable food and the sustainable sociality, arts, and crafts of the city. It has been, from the very beginning, literally a dream of paradise, in the original meaning of the Old Iranian word *paridayda* or "walled enclosure." Modern civilization has continued this dream through the idea of progress, invoking science and technology as means to assure total control over nature. But this dream has, from its inception, been a delusion. Far from controlling nature, humans have been consuming it in an unsustainable Malthusian-like trajectory, the limits of which are being breached in our time. The contraction to anthropocentric mind with the development of civilization and its later phase of the moral revolution/axial age, which I will describe below, and further contraction to mechanico-centric mind in the modern era, unnecessarily denied the enduring and ultimately sustainable conditions implied by our evolutionary legacy and tempered into the human body.

The population fed by agriculture repeatedly grew beyond limits due to the systemic conditions of agricultural settlement itself, causing a cycle of ever-increasing expansion. The social conditions invoked by agricultural civilization not only introduced monumental architecture and precise technologies, but also led to hierarchical domination, animal confinement, imperial expansion, and overwork. As civilizing humans depopulated their imaginations of wild animals and plants and replaced them progressively with human models, civilization introduced new forms of dehumanization in mass-killing warfare and slavery. Just as animals and plants had been confined through domestication, humans found new forms to confine humans within the walled enclosures of cities, even as religions increasingly began to promise otherworldly paradise.

Let us not forget the costs of civilization that are often overlooked by those not aware of the anthropological and archaeological findings. Philosopher Thomas Hobbes described the state of nature as: "the life of man, *solitary*, poor, *nasty*, brutish and short."[21] The actual state of nature, of animate mind, was one of profound sociality, and virtually the reverse of Hobbes' depiction, which was more a description of the state of civilization. With the advent of civilization, nutrition declined significantly and the human body with it. Agriculturally based civilizations universally diminished human diets by contracting the wide range of foods to one or more primary grains such as wheat, barley, or rice. Average heights dropped four to six inches in the old world as well as the new, and people worked much longer hours to sustain themselves.[22] The height increases in people of industrialized societies in the last 150 years are only a return to what they were before agriculture. Intensive agriculture and irrigation also tended to devastate landscapes—for example, those desert storms one sees in Iraq occur in what once was called "the fertile crescent."

Civilization also institutionalized the military bureaucracy and mass-killing warfare. By contrast, traditional tribal warring, though deadly, was also typically ritually bounded and did not involve mass genocide. Western Civilization—though it brutally conquered much of the world through imperial ambition and chauvinistic hubris and is a valid target of criticism—can also be seen as simply continuing the kind of dynamics found in civilizations wherever they originally formed such as in old world Mesopotamia, Egypt, the Indus Valley or China, or in the new world civilizations such as the Inca of Peru or the Aztec of Mexico.

The conditions of city living brought about the advent and proliferation of literacy, which was both a technology of increased precision in record keeping and communication, and a mind-altering instrument in the development of anthropocentric mind. Literacy was a key ingredient in the original bodying forth of bureaucratic mind. Written accounts served not only practical purposes of accounting, but also served politically legitimating and sacred purposes, as well as metaphysical outlooks. A good example of the appearance of the literate mind is the expression in Babylonian astronomy of the sky and its celestial phenomena as "the heavenly writing," embodying both a scientific astronomy and religious metaphysics. Later the "religions of the book" inscribed a deity even further removed from the earth, transcendent rather than imminent, in "heaven," as Christianity would have it, rather than in the heavens.

By the time of what John Stuart-Glennie termed "the moral revolution" and Karl Jaspers later called "the axial age" (the period roughly centered around 500–600 BCE and its legacy), the emerging world religions have stemmed from cities and are centered on human morality and transcendence from the earth. Human prophets have become the focal points, and wild nature has been either devalued or, more so in the West, is no longer an element of religion. This period marks a second phase of the development of anthropocentric mind. Religion shifts from custom to conscience, and the reflective mind takes center stage in religion, philosophy, and early science. This revolutionary period roughly 2500 years ago across civilizations brought forth new ideas exemplified by the ancient Chinese and Greek philosophers, Buddha, and Judaism (with its later offshoots of Christianity and Islam). These ideas continue to influence the way a large part of the world lives today. Most commentators on this period, the rise of what Robert Bellah called the "theoretic attitude,"[23] view it as a positive elevation of humanity, as the time when, as Jaspers said, "Man, as we know him today, came into being."[24]

Clearly, a group that includes Confucius, Buddha, Socrates and Jesus represents models of virtue and wisdom worthy of the deepest respect. Interestingly, none of these four wrote their own texts, rather their legacies depend on posthumous writings about them. Their outlooks, at least as later codified and canonized by other people in written texts, especially between 200 BCE and 200 CE, tend to center on concerns of human morality, and they are clearly exemplars of the moral revolution.[25] Yet from the perspective I am outlining here, they can also be seen as representatives of anthropocentric mind, rooted in the human point of view, contracted from the attunement to the animate earth. In place of the wisdom to be gained through the lessons of the sustaining earth, wisdom is now rooted in human philosophical and religious practices of reciprocity, self-examination, detachment, renunciation.

Socrates said that the unexamined life is not worth living, and Buddha showed the path to transcend desire and the suffering of the world through ascetic renunciation. The reflective, negating mind came to the foreground in such ideas, and that is why most commentators celebrate it. But what of the passionate, affirming mind, practically and religiously immersed in life that surrounds it, rather than idealizing a reflective detaching from it? What of the animate mind that evolved into being in love with the earth and its lessons, bodying forth in and through the philosophy of the earth, alive in the wonder? What if the acknowledgment of the wonder is primary and reflection is secondary, in the sense that humans, first and foremost, are engaged participants in the sacred game of life, and they are spectators only secondarily? That is what my diagram of the concentric circles of mind implies: that we are primally evolved as engaged and passionate participants with bounded secondary capacities for reflective, spectatorial consciousness. More simply, the affirming "Yes!" comes before the questioning "No." The undoing of that balance in the course of history, treating the secondary reflective and rationalizing mind as though primary, as though it were the basis for living, was an unsustainable mistake whose consequences are now becoming both clear and unavoidable.[26]

The religions of the book still today inform 4 billion humans on the earth and can enter into the interactions of everyday life. The global calendar is one example, informing global civilization that the date is the solar year such and such, a date based on the approximate birth of Jesus. The inscription of these histories have provided codification of beliefs, legitimized as sacred history, which are also manifestations of literate consciousness, of worldviews that metaphorically view the world in book form. In the religions of the book, for example, moral life is understood based on lessons from fixed, unrepeatable points of human history. Religious life is based on an anthropocentric outlook whose model is sacred writing, whether a celestial imprint literally, as in earlier Babylonian culture, divine signs appearing in celestial phenomena, or as the later divinely inspired sacred texts appearing in Judaic, Christian, and Muslim cultures, which, with its origins in Judaism, perhaps retained the Babylonian imprint.

David Abram has written on the disconnection from the animate earth brought about by the progressively abstract development of the alphabet in his book, *The Spell of the Sensuous*. At the Sustainable Wisdom conference he gave an example of how with 10 years of Torah study, a close student of the hidden meanings of the Hebrew words could come to realize that the name of the Creator is living breath. But think about it: Would you rather labor for ten years confined in a room as a spectator buried in a book to realize reflectively that the creator is living breath, or would you rather spend that ten years sensually immersed in that living breath as wild nature, participating in it and as it?

Abram claims that the Greek alphabet, derived from the Phoenician alphabet, lost the last vestiges of pictorial animacy, but that the Hebrew version, which was also derived from the Phoenician, still retains them. That may well be, yet Greek science maintained a learning relationship to nature continuous with Babylonian science, which is absent in Judaism, where the "moral revolution" described by Stuart-Glennie narrowed the focus of religion to the human relations with a transcendent God. The triumph of the "theoretic attitude," of greater reflectiveness came at the cost of perceptive learning relationships with the living habitat as a centerpiece of religion, a process already begun with the earlier civilizational religions.

Religion became, less so in the East but especially so in the West, deranged from the creating cosmos, which is neither a God concept nor life-transcending Nirvana, nor inscribed sacred history, but the living spontaneity of psycho-spiritual-physical ongoing creation. The limited human mind, now anthropocentrically fixed on its concept of the ultimate as a transcendent mind quality, seeks to control nature instead of learning from it, especially in the western variants of the religions of the book, but also in Greek science.

Even though the traditions stemming from ancient Greek science took nature as a source of learning, knowledge (or *epistêmê*) gave way to technique (*technê*) in the rise of modern science and technology. To control the cosmos, through supernatural religious mind or technical knowledge, as though the cosmos could be reduced to mind, is the drama that took root some 2,500 years ago. Yet the cosmos cannot be controlled, but only participated in: for it is the primal drama of creation, the oldest drama out of which, through affirmative attunement to the wonder, humanity and human religion emerged.

The Mechanical Worldview as Contraction

If civilization was the advent of anthropocentric mind, contracted from immersion in the wild habitat to a domesticated one marked by the figure of the human in buildings and beliefs, then the modern world represents a new phase of contracted consciousness, that of "mechanico-centric mind" as dominant worldview, and centered in the rational mechanical elements of mind and projections which represented them, beginning with the clock. Though manifest originally in the largely human bureaucratic machine that was civilization, a new configuration re-appeared in the late medieval period that would heighten the centrality of the non-human elements of the machine, exhibited first in the invention of the mechanical clock.

We might speak of a kind of crypto- "divine birth" in some Northern European Benedictine Monastery in the 1270s: the appearance of the mechanical clock, which soon spread both as machine and as symbol. It was first developed as an aid to rationally ordering the seven daily prayer times in the monastery, later diffused into towns and workplaces. Max Weber suggests in his book, *The Protestant Ethic and the Spirit of Capitalism*, that it was as if with the advent of the reformation the virtue of ascetic rationality within the doors of monastery had been let loose, and everyone had to practice ascetic rationality. Taking off from that metaphor, we might say that it was the mechanical clock and its discipline of uniform rational precision that exited those monastery walls.

The clock expanded from being a marker of prayer time in the monastery, to the town clock, to the ubiquitous regulator of time, work discipline, and capitalism.[27] By the 17th century, clock culture had grown such that the great astronomer Kepler could say, "I do not see the heavens as a divine, live being but as a kind of clockwork."[28] Scientific materialism was born of the clock, and remains encapsulated by its machine-as-metaphor legacy. Descartes and Hobbes each saw the human body as an apparatus. The machine, as a precise, though limited, projection of the rational portion of mind, had come to define mind itself. We developed a clockwork universe, incredibly vast, yet whose quantification of time and space comes with costs: The universe for our foraging ancestors was alive, and alive with wonder and meaning. For us, it is inanimate clockwork, or some updated machine metaphor, to which we are reducible, or, euphemistically, "uploadable."

The objectivist machine metaphor was a powerful factor in the massive spread of the machine in modern life, in precision renderings of the world picture. But the world is far more than the limited metaphor allows it to be. The machine metaphor turns out to be a form of confinement. The great progress in precision came at the cost of cutting off other realities which did not fit its precise but limited perspective, as well as removing ourselves from them: the "milk of human kindness," of capacities for empathic relation and parenting, for deep awareness, spontaneously alive to the present, for imaginative projection and poetic wonder, for the palpable touch of organic life. These are the still living legacies of our evolution into humankind, too often denigrated and devalued by the rational-mechanical bureaucratic mindset.

The modern worldview of the machine, and earlier human-centric worldview of civilization, emerged out of a vast prehistoric legacy vastly different, yet also still powerfully significant today. The waves of dispersals of earlier humans, and finally anatomically modern humans, over the globe found their way into various habitats that required attunement to the specifics of the local habitat. But in all these ways of living there remained the constant of foraging and the common elements it required, which were continuous learning relations through attunement to habitat in practical and reverential ways, as well as progressively inclusive sociality, from parenting to foraging to ritual life. And, it is out of these common elements that the very capacity for *symboling*, for communicating through symbolic signs in ritual, language, and art, originally emerged.

Consider the implications of these three phases of consciousness. The living cosmos embodied as the animate earth is one of ongoing creation in the perspective of animate mind, involving being attuned to the living earth as participants: a participation consciousness. As a Wemindji Cree Native Canadian man quoted by anthropologist Tim Ingold put it, life is "continuous birth," and in Ingold's words, "incipient, forever on the verge of the actual … One is continually present as witness to that moment, always moving like the crest of a wave, at which the world is about to disclose itself for what it is."[29] It's like riding a wave, immersed in the flow that is ever bodying forth; it is ongoing creation. So, your outlook is that of immersion in a living process, in which I am an active participant in the ongoing wave of insurgent life.

With the contraction to "anthropocentric mind," the world becomes a spectacle: the spectacle of the palace, the spectacle of the walled city, the spectacle of the new divine kingship and its rigid political and religious hierarchies. And I? Who am I in that? If I am not the king I am a spectator of the great drama of civilization with its sacred kings, prophets, and texts. Religion turns from the attunement to the living wild other in ongoing creation, to texts set in a fixed historical past that mark the sacred: religions of the book.

With the advent of the modern mechanico-centric era, the universe is a machine, contracted to the projection of the automatic functions of the human psyche. And who am I in that? I am but the ghost in the machine, that Cartesian dualistic separation of spectral mental consciousness from matter. Again, the mechanico-centric mind and its scientific-technical outlooks provided precise abstract accounts of the world, contributing many positive achievements to modern life, from improved transportation, utilities, and communication to the multitude of scientific discoveries. It opened up unprecedented powers, but often without knowing what their sustainable limits should be. It assumed that its abstract map was the actual concrete landscape.

The contractions to anthropocentric and mechanico-centric mind were accompanied by the sense that they were progressive expansions, upward and onward, for the good of humanity. With hindsight we can begin to see how they were instead the ballooning expansion of the human ego, boundlessly seeking escape from the earth. We went from a philosophy of the earth and animate mind—our long legacy of foraging—to a philosophy, ever since civilization began, of escape from the earth. And today, for the very first time since civilization began, we are presented with the dilemma of how to come up with limits across the board which can offset that murderous and suicidal global destruction now underway.

Conclusion: Refinding the Philosophy of the Earth

Upon the vast, incomprehensible pattern of some primal morality greater than ever the human mind can grasp, is drawn the little, pathetic pattern of man's moral life and struggle, pathetic, almost ridiculous.

—D. H. Lawrence[30]

Far from the "up, up, and away" story of unfettered progress, also known as our unsustainable world, the loss of the touch of the earth was part of the *progressive confinement of mind*. Unfettered progress has come with devastating consequences to *panzoa* (all life), increasingly manifest in dwindling wildlife, acidifying oceans, dying corals, shrinking resources of water, and many other indicators as we careen into a cascading manifestation of suicidal unsustainability.

In the modern era, technology has been invoked as a partner in the rise of democracy and means to new freedom and leisure, yet it was also a collaborator in genocidal imperialism, industrial dehumanization, modern totalitarianism, and capitalizing global biocide. The myth of the machine today, specifically of its scientific and technological institutions and ideological elements, is that its continued expansion will automatically be good for humanity, and that its continued development will inherently provide solutions, as though on their own rather than as projections of human purposes and human prejudices.

The attempt to control nature was, in effect, an attempt to transcend the laws of creation. Consider the words of "Darwin's bulldog," as he was known, the early advocate of Darwin's theory of evolution, T. H. Huxley:

Let us understand, once for all, that the ethical progress of society depends, not on imitating the cosmic process, still less in running away from it, but in combating it. It may seem an audacious proposal thus to pit the microcosm against the macrocosm and to set man to subdue nature to his higher ends; but I venture to think that the great intellectual difference between the ancient times with which we have been occupied and our day, lies in the solid foundation we have acquired for the hope that such an enterprise may meet with a certain measure of success.[31]

It has become increasingly obvious that the combating of cosmos has not been an evolutionary advance, but rather an ongoing march toward disaster. It has now become evident that the dream of transcendence, of paradisiacal escape from the earth through technology or religion, was delusional; it was simply a violation of the laws of creation. The clash against the cosmos, depicted as the magnificent march of progress, speaks of the overweening pride of the unmatured primate, glorying in the power of its "higher" brain functions.

Contracted to anthropocentrism with the advent of civilization and later phase of the moral revolution, contracted even further in the modern phase of mechanico-centrism, modernizing humanity naively believed that its upper brain, and its rational mechanical instruments and institutions, its science and technology, its materialistic economics, blessed in many cases by its etherealized religions, was sufficient to overpower nature. But, the animate earth is far deeper, richer, more sustaining, tempered, and complex than that callow mindset could comprehend.

The contractions of mind brought about through the rise of mechanico-centric mind in the modern era, and of anthropocentric mind in the rise of civilization and its second phase of the moral revolution, represent apparent progress in rational precision. But, simultaneously they are regressive in that they disregard the realities of the wild earth, in terms of limits as well as modalities of awareness, in terms of subjective experience, and in bodying forth participation mind, or what D. H. Lawrence called "affirmative mind," as insignificant. What was glorified as expansive "progress" actually involved a progressive confinement, the loss of the touch of the earth. The contractions of mind outline not simply the progress in forms of technical precision, but also the regressive retreat from reality that came to picture the world as a schizoid, feelingless machine.

The escape from the earth has always promised liberation, many times attaining it in the short run, only to lose it in the long run that has begun to come better into focus. Though the modern outlook, through science, has been coming around to better appreciate the sustainable outlooks of first ways peoples in terms of sustainable practices better rooted in ecological mind, there remains a disconnect when it comes to the possibilities of sustainable wisdom. What those peoples do through sustainable practices may look good, but what they believe remains suspect to the modern mind.

The maximizing ideology of the machine which dominates the modern worldview is not only suicidally unsustainable generally (which is, from another point of view, a dark kind of definition of the successful escape from the earth), but is an inadequate framework for science itself, as thinkers such as philosopher and practicing scientist Peirce and also Mumford argued. It is as though science wants to let the facts speak, but because it is still dominated by the myth of the machine, by so-called "scientific materialism" that is not adequate to the broader vision required by science, it is not developed sufficiently to allow the voice of the earth to speak through facts not format-able to the grid of the machine. We never left the earliest sphere of animate mind attuned to animate earth, but only denied it, and real progress may involve realizing our original evolutionary relationship to the earth still holds the tempered wisdom through which to reimage a sustainable earth. If so, a renewed philosophy of the earth, honed from the wisdom imparted by the earth as primal model rather than the abstraction of the machine, may yet speak to the further developments of science and society.

What then is the relationship of the animate earth, as source of animate mind, to science and to the greater universe in general? In brief, animate mind provides a means to open the contracted mechanico-centric worldview and its science to a broader conception of the universe as living creation, a conception of an *animate universe* that must yet do justice to the requirements of science. Such an outlook is suggested by the unlikely source of arch-logician and scientist Peirce, who already showed over a century ago precisely a rigorous, yet broadened conception of science that in effect reactivates the missing ingredients of animate mind.[32]

Peirce rejected the scientific worldview of necessitarianism, of the clockwork universe moving with the necessity of a clock. He showed the irreducible element of spontaneity in the development of variescence, something scientific positivism then and now could not account for, and argued for the continuing evolution of laws, including physical laws. And, in the further development of his semiotic realism, he challenged science to come to terms with a more comprehensive living universe, alive in still active creation and a reasonableness energizing into being.

Peirce's scientific theory of reality involved a kind of logical extrapolation of "seven generations" thinking, to an "unlimited community of inquiry" capable of self-corrective learning into the horizon of the future, a theory ineradicably rooted in a broad social principle not limited to humans:

> It seems to me that we are driven to this, that logicality inexorably requires that our interests shall not be limited. They must not stop at our own fate, but must embrace the whole community. This community, again, must not be limited, but must extend to all races of beings with whom we can come into immediate or mediate intellectual relation. It must reach, however vaguely, beyond this geological epoch, beyond all bounds. He who would not sacrifice his own soul to save the whole world, is, as it seems to me, illogical in all his inferences, collectively. Logic is rooted in the social principle.[33]

Panzooinism, the ongoing livingness of things, as the source and legacy of sustainable wisdom, may yet have something to say, not as an inadequately framed conception of nature in reductionist versions of science and technology, but as a guidepost to what it means to re-attune contemporary civilization to the laws, limits, and poetic wonder of the variescent earth. The task, as I see it, is not to jettison the legacies of the modern mechanico-centric mind and older anthropo-centric mind, including the lessons of the moral revolution and the modern scientific outlook, but to *optimize* them by returning them to their limited place in the larger framework of animate mind attuned to the animate earth. The false myth of history as "up, up, and away" linear progress reveals a more contradictory history as a paradoxical progressive regression, whose contractions have masked what was there all along. Although it remains as an eclipsed reality today, it begins to come into view: the relation to the sustaining properties of the earth, with its limits and possibilities, is the key to long-term human flourishing.

Optimistic Epilogue

So let's be optimistic, like the legendary climatologist and former NASA scientist James Lovelock, the founder of "the Gaia hypothesis," the scientific idea that the earth is a self-regulating homeostatic system. When asked over 50 years ago, in 1965, by Shell Oil Company what the year 2000 would be like, he broke from the optimistic "up, up, and away" futuristic accounts of technology given by other experts and told Shell that the environment "will be worsening then to such an extent that it will seriously affect their business."[34] His concerns with the environment have continued into the present. In an interview in 2008, he predicted things would be getting very bad in 20 years (by 2028):

> There have been seven disasters since humans came on the earth, very similar to the one that's just about to happen. I think these events keep separating the wheat from the chaff. And, eventually we'll have a human on the planet that really does understand it and can live with it properly. That's the source of my optimism.[35]

That human who really does understand it and can live with it properly is not simply a phantasm of the future. It is the still living legacy of those who live and have lived by the philosophy of the earth, aware of the need for self-controlling, sustainable limits on one's culture at all levels of institutions and beliefs, as well as that legacy still living in our evolved Pleistocene bodies today. It is the realization that the earth was not put here for humans, as the philosophy of the escape from the earth believes, but that it is a great and marvelous gift out of which humans bodied forth to serve its continued flourishing. This is the worldview we honored for most of human history, and it is the legacy that surviving Indigenous cultures can still teach us.

Civilization, in my opinion, has already begun the cascade to collapse that Lovelock described, one that a 2014 NASA research project depicted in similar dark outlines, illustrating the role of economic inequality in the decline.[36] Technology alone, the *Deus ex Machina* "god out of the machine," will not save us at this late date. I do not think it likely that global civilization can sufficiently awaken technically, mentally, or spiritually to face the looming catastrophic consequences now taking shape of the tragic delusion of escape from the earth. Quite the opposite, I am sad to say. But I am sure that animate mind, inclusive of the more limited outlooks which grew out of balance in the course of civilization, will continue to nurture the human spirit of those attuned to it. There is good reason to hope in the sustainable wisdom of the earth, or, as Harry Roberts put it, "the earth pushing back, supporting us … the earth loving us."

Discussion Questions

What is the philosophy of the Earth? How does the advent of agriculture and civilization represent a departure from the philosophy of the earth?

What is meant by the contrast between "the philosophy of the earth" and "the philosophy of escape from the earth?"

What does the idea of history as a "contraction of mind" rather than simple progress mean?

What do "animate mind," "anthropocentric mind," and "mechanico-centric mind" mean?

Notes

1. Laura Burges, "Jokan Zenshin Thomas (Tim) Buckley: Zen Way, Yurok Way," (2015) San Francisco Zen Center Website, accessed 10 July 2016, http://blogs.sfzc.org/blog/2015/04/27/zen-way-yurok-way-republished/.

2. See Eugene Halton, "Planet of the Degenerate Monkeys," in *Planet of the Apes and Philosophy*, ed. John Huss (Chicago, IL: Open Court Press, 2013), 279–292. See also my "Eden Inverted: On the Wild Self and the Contraction of Consciousness," *The Trumpeter* 23, no. 3 (2007): 45–77. For a discussion of panzoonism, see my *From the Axial Age to the Moral Revolution: John Stuart-Glennie, Karl Jaspers, and a New Understanding of the Idea* (New

York, NY: Palgrave MacMillan, 2014). In that book I resurrect the forgotten philosophy of history of folklorist John Stuart-Glennie, who was the first to develop a theory of the period of roughly 600 BCE when many of the world religions still present came into being from across different civilizations, a period he termed the moral revolution. Stuart-Glennie distinguished his theory of panzoonism from E. B. Tylor's theory of animism, claiming that Tylor's term was misleading, implying a spirit inhabiting a thing from without, rather than the living power inherent in things themselves.

3. See also David Lavery, *Up From the Skies*, (Carbondale, IL: Southern Illinois University Press, 1992), available online at: http://davidlavery.net/LFS/.

4. Ray Kurzweil, *The Singularity is Near: When Humans Transcend Biology* (New York, NY: Penguin Group, 2005), 9.

5. Peirce described variescence as follows: "I may however spend a few minutes in explaining a bit more clearly what I mean by saying that if the universe were governed by immutable law there could be no progress. In place of the word progress I will put a word invented to express what I mean, to wit, such a change as to produce an uncompensated increment in the number of independent elements of a situation." Charles S. Peirce, *Semiotic & Significs: The Correspondence Between Charles S. Peirce & Victoria Lady Welby*, ed. Charles Hardwick (Bloomington, IN: Indiana University Press, 1977), 143.

6. Henry Adams and Brook Adams, *The Degradation of the Democratic Dogma* (New York, NY: Macmillan, 1919), 267–311.

7. See Lewis Mumford, *The Myth of the Machine: Technics and Human Development*, vol.1 (New York, NY: Harcourt Brace, 1967); and *The Pentagon of Power*, vol.2 (New York, NY: Harcourt Brace, 1970).

8. This outlook derives from the legacy of the classic nominalist philosophy that emerged from the late medieval period, which at the time was termed "the modern way" (*via moderna*), and which, indeed, came to dominate modern philosophy and science. It proposed a divided view, of nature on the one side as isolate particulars devoid of general relations versus sociality on the other as blank conventions, concealing the interactive realities of nature and human sociality, not to mention the natural basis of sociality more generally. Actual human development is a process of "nature-nurture," wherein inborn characteristics emerge under proper conditions of parenting, truly biological and truly social.

9. Eugene Halton, "Mind Matters," *Symbolic Interaction* 31, no. 2 (2008): 119–141.

10. Paul Shepard, *The Tender Carnivore and the Sacred Game* (Athens, GA: University of Georgia Press, 1998 [1973]), 6.

11. Shepard, *The Tender Carnivore and the Sacred Game*; Nurit Bird-David, "'Animism' Revisited: Personhood, Environment, Relational Epistemology," *Current Anthropology* 40, No. Suppl. 1 (1999): S67–S91; Steven Mithen, "The Hunter-Gatherer Prehistory of Human-Animal Interactions," *Anthrozoös*, 12, no. 4 (1999): 195–204.

12. Robin Wall Kimmerer, "Learning the Grammar of Animacy," in her *Braiding Sweetgrass: Indigenous Wisdom, Scientific Knowledge and the Teachings of Plants* (Minneapolis, MN: Milkweed Editions, 2015), 48, 57–58.

13. Halton, "Eden Inverted"; "Planet of the Degenerate Monkeys"; and "From the Emergent Drama of Interpretation to Enscreenment," in *Ancestral Landscapes in Human Evolution: Culture, Childrearing and Social Wellbeing*, ed. Darcia Narvaez, Kristin Valentino, Agustin

Fuentes, James McKenna, and Peter Gray (Oxford, UK: Oxford University Press, 2014), 307–330.

14. Charles Peirce, *Preface to Essays on Meaning* [unpublished manuscript], 23 Oct 1909, MS 640, Widener Library, Harvard University.

15. In a 1906 discussion of what he meant by *quasi-mind*, Peirce claimed that, "Thought is not necessarily connected with a brain. It appears in the work of bees, of crystals, and throughout the purely physical world; and one can no more deny that it is really there, than that the colors, the shapes, etc., of objects are really there ... Not only is thought in the organic world, but it develops there." See Charles Peirce, "Prolegomena to an Apology for Pragmaticism," *Collected Papers* Vol. 4, Para. 551 (Cambridge, MA: Harvard University Press, 1933).

16. William Blake, from his poem, *Jerusalem* 3, E 146. Accessed online 15 May 2016 at Bartleby.com, https://www.bartleby.com/235/304.html.

17. Halton, "Eden Inverted"; *From the Axial Age to the Moral Revolution*; "From the Emergent Drama of Interpretation to Enscreenment."

18. Initially, my diagram did not picture the animate earth as the outer layer, but following a presentation, biologist Lynn Margulis (associated with the Gaia hypothesis), enthusiastically insisted that I include animate earth in the picture. She shoved a book into my hands for which she had written the introduction, it was Stephan Harding's *Animate Earth*. I was further delighted (later, at the sustainable wisdom conference) to discover that Harding's title had been suggested by conference participant David Abram. See Stephan Harding, *Animate Earth: Science, Intuition, and Gaia*, (White River Junction, VT: Chelsea Green, 2006).

19. Ian Hodder, interviewed in "From Hunter-Gatherer to Farmer (Part 2)," accessed online 4 August 2014, https://www.youtube.com/watch?v=iDxDPvkzr1s.

20. Klaus Schmidt, "Zuerst kam der Tempel, dann die Stadt, Vorläufiger Bericht zu den Grabungen am Göbekli Tepe und am Gürcütepe 1995–1999," *Istanbuler Mitteilungen* 50 (2000): 5–41.

21. Thomas Hobbes, *Leviathan*, Pt. I, Ch. XIII. The Project Gutenberg EBook of Leviathan. Accessed 23 January 2017, https://www.gutenberg.org/files/3207/3207-h/3207-h.htm.

22. Stanley Boyd Eaton, Marjorie Shostack, and Melvin Konner, *The Paleolithic Prescription: A Program of Diet and Exercise and a Design for Living* (New York, NY: Harper & Row, 1988); Richard Lee and Irven DeVore, *Man the Hunter* (Chicago, IL: Aldine, 1968); Amanda. Mummert, et al., "Stature and Robusticity During the Agricultural Transition: Evidence from the Bioarchaeological Record," *Economics and Human Biology* 9, no. 3 (2011): 284–301; Marshall Sahlins, *Stone-Age Economics* (Chicago, IL: Aldine Transaction, 1973).

23. Robert Bellah, *Religion in Human Evolution: From the Paleolithic to the Axial Age* (Cambridge: Belknap Press of Harvard University Press, 2011), 275.

24. Karl Jaspers, *The Origin and Goal of History* (New Haven: Yale University Press, 1953), 2.

25. Jan Assmann, *Cultural Memory and Early Civilization: Writing, Remembrance, and Political Imagination* (Cambridge, UK: Cambridge University Press, 2012), 399.

26. Eugene Halton, "Wonder, Reflection, and the Affirmative Mind: From Panzoonism to Apocalypse," (paper presented at the conference on Wonder and the Natural World, Indiana University, Bloomington, IN 16 June 2016); and "The Forgotten Earth: World Religions and Worldlessness in the Legacy of the Axial Age/Moral Revolution," in *From World*

Religions to Axial Civilizations, ed. Said Arjomand and Stephen Kalberg. (Submitted for publication, 2018).

27. E. P. Thompson, "Time Work-Discipline, and Industrial Capitalism," *Past & Present* 38, no. 1 (1967): 56–97.

28. Johannes Kepler, "Letter to J. G. Herwart von Hohenberg," 16 February 1605, quoted in Arthur Koestler, "Johannes Kepler," *Encyclopedia of Philosophy* 4 (New York: Macmillan, 1967), 331.

29. Tim Ingold, *Being Alive: Essays on Movement, Knowledge and Description* (New York, NY: Routledge, 2011), 69.

30. D. H. Lawrence, *Phoenix*, edited and introduced by Edward D. McDonald (New York, NY: Viking, 1936), 419.

31. Thomas Henry Huxley, *Evolution and Ethics and Other Essays* (1893; reprint, New York, NY: Barnes and Noble, 2006), 49.

32. Eugene Halton, "The Transilluminated Vision of Charles Peirce," in *Bereft of Reason: On the Decline of Social Thought and Prospects for Its Renewal* (Chicago, IL: University of Chicago Press, 1995), 167–89.

33. Charles Peirce, *Essential Peirce: Selected Philosophical Writing*, vol. 1 (1867–1893), edited by Nathan Houser and Christian Kloesel (Indianapolis, IN: Indiana University Press, 1992), 149.

34. Decca Aitkenhead, "James Lovelock: 'Enjoy Life While You Can: In 20 Years Global Warming will Hit the Fan,'" *The Guardian*, 29 February 2008, https://www.theguardian.com/theguardian/2008/mar/01/scienceofclimatechange.climatechange.

35. Ibid.

36. Safa Motesharrei, Jorge Rivas, Eugenia Kalnay, "Human and Nature Dynamics (HANDY): Modeling Inequality and Use of Resources in the Collapse or Sustainability of Societies," *Ecological Economics* 101 (2014): 90–102.

References

Adams, Henry. "The Rule of Phase Applied to History." In *The Degradation of the Democratic Dogma*, edited by Henry Adams and Brook Adams, 267–311. New York, NY: Macmillan, 1919.

Aitkenhead, Decca. "James Lovelock: 'Enjoy Life While You Can: In 20 Years Global Warming will Hit the Fan.'" *The Guardian*, 29 February 2008. https://www.theguardian.com/theguardian/2008/mar/01/scienceofclimatechange.climatechange

Assmann, Jan. *Cultural Memory and Early Civilization: Writing, Remembrance, and Political Imagination*. Cambridge, UK: Cambridge University Press, 2012.

Bellah, Robert. *Religion in Human Evolution: From the Paleolithic to the Axial Age*. Cambridge: Belknap Press of Harvard University Press, 2011.

Bird-David, Nurit. "'Animism' Revisited: Personhood, Environment, Relational Epistemology." *Current Anthropology* 40, no. Supp. 1 (1990): S67–S91.

Burges, Laura. "Jokan Zenshin Thomas (Tim) Buckley: Zen Way, Yurok Way." (April 27, 2015) San Francisco Zen Center Website. – Accessed 10 July 2016. http://blogs.sfzc.org/blog/2015/04/27/zen-way-yurok-way-republished/.

Eaton, Stanley Boyd, Marjorie Shostack, and Melvin Konner. *The Paleolithic Prescription: A Program of Diet and Exercise and a Design for Living.* New York, NY: Harper & Row, 1988.

Halton, Eugene. "Eden Inverted: On the Wild Self and the Contraction of Consciousness." *The Trumpeter* 23, no. 3 (2007): 45–77. Also available online http://trumpeter.athabascau.ca/index.php/trumpet/article/view/995.

———. *From the Axial Age to the Moral Revolution: John Stuart-Glennie, Karl Jaspers, and a New Understanding of the Idea.* New York, NY: Palgrave MacMillan, 2014.

———. "From the Emergent Drama of Interpretation to Enscreenment." In *Ancestral Landscapes in Human Evolution: Culture, Childrearing and Social Wellbeing, edited* by Darcia Narvaez, Kristin Valentino, Agustin Fuentes, James McKenna, and Peter Gray, 307–330. Oxford, UK: Oxford University Press, 2014.

———. "Mind Matters." *Symbolic Interaction* 31, no. 2 (2008): 119–141.

———. "Planet of the Degenerate Monkeys." In *Planet of the Apes and Philosophy*, edited by John Huss, 279–292. Chicago, IL: Open Court Press, 2013.

———. "The Forgotten Earth: World Religions and Worldlessness in the Legacy of the Axial Age/Moral Revolution." In *From World Religions to Axial Civilizations*, edited by Said Arjomand and Stephen Kalberg. Submitted for publication, 2018.

———. "The Transilluminated Vision of Charles Peirce." In *Bereft of Reason: On the Decline of Social Thought and Prospects for Its Renewal*, 167–89. Chicago, IL: University of Chicago Press, 1995.

———. "Wonder, Reflection, and the Affirmative Mind: From Panzoonism to Apocalypse." Paper presented at the conference on Wonder and the Natural World. Indiana University, Bloomington, IN, June 2016.

Harding, Stephan. *Animate Earth: Science, Intuition, and Gaia.* White River Junction, VT: Chelsea Green, 2006.

Hobbes, Thomas. *Leviathan*, Pt. I, Ch. XIII. The Project Gutenberg EBook of Leviathan. Accessed 23 January– 2017, https://www.gutenberg.org/files/3207/3207-h/3207-h.htm.

Hodder, I. Interviewed in *From Hunter-Gatherer to Farmer* (Part 2). Accessed 4 August 2014. https://www.youtube.com/watch?v=iDxDPvkzr1s

Huxley, Thomas Henry. *Evolution and Ethics and Other Essays.* 1893. Reprint. New York, NY: Barnes and Noble, 2006.

Ingold, Tim. *Being Alive: Essays on Movement, Knowledge and Description.* New York, NY: Routledge, 2011.

Jaspers, Karl. *The Origin and Goal of History.* New Haven: Yale University Press, 1953.

Kepler, Johannes. "Letter to J. G. Herwart von Hohenberg." 16 February 1605. Quoted in Arthur Koestler. "Johannes Kepler." *Encyclopedia of Philosophy* 4, 331. New York: Macmillan, 1967.

Kimmerer, Robin Wall. "Learning the Grammar of Animacy." In her *Braiding Sweetgrass: Indigenous Wisdom, Scientific Knowledge and the Teachings of Plants*, 48–58. Minneapolis: Milkweed Editions, 2015.

Kurzweil, Ray. *The Singularity Is Near: When Humans Transcend Biology*. New York, NY: Penguin Group, 2005.

Lavery, David. *Up From the Skies*. Carbondale, IL: Southern Illinois University Press, 1992. Available online at http://davidlavery.net/LFS/

Lawrence, D. H. *Phoenix*. Edited and introduced by Edward D. McDonald. New York, NY: Viking, 1936.

Lee, Richard, and Irven DeVore. *Man the Hunter*. Chicago, IL: Aldine, 1968.

Mithen, Steven. "The Hunter-Gatherer Prehistory of Human-Animal Interactions." *Anthrozoös*, 12, no. 4 (1999): 195–204.

Motesharrei, Safa, Jorge Rivas, and Eugenia Kalnay. "Human and Nature Dynamics (HANDY): Modeling Inequality and Use of Resources in the Collapse or Sustainability of Societies." *Ecological Economics* 101 (2014): 90–102.

Mumford, Lewis. *The Myth of the Machine: Technics and Human Development*, vol. 1. New York, NY: Harcourt Brace, 1967.

_____. *The Myth of the Machine: The Pentagon of Power*, vol.2. New York, NY: Harcourt Brace, 1970.

Mummert, Amanda, Emily Esche, Joshua Robinson, and George J. Armelagos. "Stature and Robusticity During the Agricultural Transition: Evidence from the Bioarchaeological Record." *Economics and Human Biology* 9, no. 3 (2011): 284–301.

Peirce, Charles. *Essential Peirce: Selected Philosophical Writing*, vol.1 (1867–1893). Edited by Nathan Houser and Christian Kloesel. Indianapolis, IN: Indiana University Press, 1992.

_____. *Preface to "Essays on Meaning"* [unpublished manuscript]. 23 Oct 1909, MS 640. Cambridge, MA: Widener Library, Harvard University.

_____. "Prolegomena to an Apology for Pragmaticism." *Collected Papers* Vol.4, Para. 551. Cambridge, MA: Harvard University Press, 1933.

_____. *Semiotic & Significs: The Correspondence Between Charles S. Peirce & Victoria Lady Welby*. Edited by Charles Hardwick. Bloomington, IN: Indiana University Press, 1977.

Sahlins, Marchall. *Stone-Age Economics*. Chicago, IL: Aldine Transaction, 1973.

Schmidt, Klaus. "Zuerst kam der Tempel, dann die Stadt. Vorläufiger Bericht zu den Grabungen am Göbekli Tepe und am Gürcütepe 1995–1999." *Istanbuler Mitteilungen* 50 (2000): 5–41.

Shepard, Paul. *The Tender Carnivore and the Sacred Game*. Athens, GA: University of Georgia Press, 1998 [1973].

Thompson, E. P. "Time Work-Discipline, and Industrial Capitalism." *Past & Present* 38, no. 1 (1967): 56–97.

CHAPTER FOUR

Rematriating Economics

The Gift Economy of Woodlands Matriarchies

BARBARA ALICE MANN

Western anthropology has finally, if begrudgingly, acknowledged the existence of Indigenous North American matriarchies and, moreover, not as an organizing principle of "primitive" or "pre-industrial" cultures but as a fully modern form of self-governance, in no way subordinate or preliminary to statism. Euro-scholarship still lags, however, in recognizing that the second half of Western statism, its exchange economy, remains a pure expression of the European raiding culture.[1] The alternative to both statism and raiding is found in the sophisticated structures of governance of the Indigenous North American matriarchies, as supported by their intricately interwoven distribution system of the gift economy. Neither can the gift economy be separated from Indigenous matriarchies, for globally, the one, steady feature of all matriarchies is some permutation of the gift economy, as I, along with others have been noting since at least 1999.[2]

Until embarrassingly recent times, Western statism was pressed as the *only* modern mode of governance, with offended Euro-scholars not so much discussing, as heatedly denouncing, the very notion that culture had any alternative models. This is a stance of long-standing. From Hippocrates in 400 BCE with his breast-mutilating Scythian mothers to mid-twentieth-century sociologists denigrating matriarchy as "the sheerest fantasy," impartial inquiry into matriarchy and its gift economy has been effectively drowned out.[3] On the one hand, Euro-scholars insisted that any non-statist system must necessarily be gobbled up by states and empires, while on the other hand, they contended that gift economies were such crude systems of barter that they would naturally yield to the pressure of the cultured system of capitalism.[4] In the late twentieth century, it was not unusual for Western scholars even to claim that the gift economy was no economy at all, while some grumblers still seek to denigrate gift economies as con games, which, presenting a seeming reciprocity, actually work as a form of disguised extortion.[5]

Western feminists have eagerly participated in the naysaying. The gag reflex for them centered, and continues to center, on matriarchy's emphasis on motherhood, birthing, and child-rearing. Apparently, any talk of motherhood is a patriarchal plot to land feminists back in the kitchen, barefoot and pregnant. Consequently, the mid-twentieth-century feminist icon Simone de Beauvoir equated mothering with slavery, yet it was de Beauvoir who entertained patriarchal myth in dutifully reciting anthropology's one-size-fits-all "stages of history."[6] This particular justification of nineteenth-century colonial conquest and exploitation fantasized that all societies, everywhere, throughout all time had developed precisely as European society supposedly had, lurching ever upwards from free-for-all matriarchy to patriarchy, through the bowels of slavery-requiring agriculture.[7] Under this construction, culture was a twelve-step program in which motherhood was insufficient as a status to allow women "to conquer the highest rank."[8] Betty Friedan dolloped off this Euro-feminist distaste for biological motherhood in her 1963 *Feminine Mystique* by recommending that middle-class women escape their children, leaving uneducated, low-income, and presumably minority women to take on the "drudgery" of childcare, the better to release deserving, upscale Euro-professional women to the intellectual excitement of career-building.[9]

Rounding the curve into the twenty-first century, to disguise such obvious elitism, straw-man-style argumentation crept to the fore of feminist ideology, especially in Cynthia Eller's didactic declaration that *any* claim of matriarchy was delusional.[10] Western feminists urgently denounced Indigenous American matriarchy as "primitive" and "unsustainable," while exalting their politics of self-interest against Indigenous critiques of capitalism, pretending that all such reappraisals flowed either from "discredited" socialism or from racist identity politics. Styling themselves as warriors of "anti-essentialism," they now presented their fight as a noble one against racist imperatives through a proper grasp of the law—despite the fact, as Cheikh Anta Diop and Ifi Amadiume have amply shown, that Western "juridical" (legal) arguments are intensively skewed to the patriarchal.[11] Happily today, Western elites can no longer dictate the terms of every discussion, so others have begun talking back, forcefully demonstrating that matriarchies are neither "primitive," nor "mythical," nor "unsustainable."[12]

Against these partisan attacks, early Western theorists of the elegant and sustainable gift, most especially Genevieve Vaughan, argued that the gift economy was foundational to the capital-accumulating raid. Without the pre-existence of a gift-economy, she held, capital raiding and forced exchange could not possibly survive. Presenting the capitalist economics of exchange ("*homo economicus*") as, essentially, a parasite latching onto the healthy body of the gift economy, Vaughan argued that the exchange parasite fed on the gifting host ("*homo donans*") until the host collapsed, taking down both systems.[13] In an interesting parallel of Steven Newcomb's insight that European linguistic noun structures created philosophical mindsets that were then elevated by Western theorists into global, spiritual truths for use as legal caveats (as found in his 2008 *Pagans in the Promised Land*), Vaughan, in her 1997 work posited that economics was a linguistically derived "mindset of exchange," spinning out its own purportedly unchallengeable truths.[14] In her thesis, Vaughan proposed Western economic meaning as having been created by the interplay of transitive (gift economy) and intransitive (capital economy) verb phrases.[15] Brilliantly original, these twin treatises offer powerful tools for analyzing the formation of theoretical models of governance and economics which, now alert to, can be free of unexamined Western imperatives.

Specifically, the two foundational concepts of Indigeneity—matriarchy and its gift economy—can now be propounded on their own terms. The first piece in play is the Indigenous governance structure, which differs dramatically from the European model, from which, to date, all "mainstream" discourse has proceeded. Whereas Europe and its derived cultures operated, and still operate, on the centralized model of hierarchy that yields statist control, Indigenous North American cultures work/ed from the matriarchal model in either decentralized confederacies or in purely distributive communities. The three models can be visualized schematically, as in Figure 4.1, below.

Western statist structures are temporal in concept and typically patriarchal, with all power residing in a center that is maintained by force.[16] The early feudal center expands, progressively engulfing all peripheral resources to hoard them for its elites. Once all resources within the state's purview have been tapped, either the state collapses as the periphery now comes at the center (think: Ancient Rome), or it begins violently expanding its periphery by raiding new groups (think: both British Empires). Usually, the first imperial raids begin as one-off, hit-and-run events, like, say, those under the sixteenth-century British pirate, Sir Francis Drake. If, however, the preliminary raids go safely enough, raided lands are secured imperially in permanent structures as "colonies," as were the Caribbean Leeward Islands after raids of Drake and his contemporary ilk.[17]

Three Governance Models

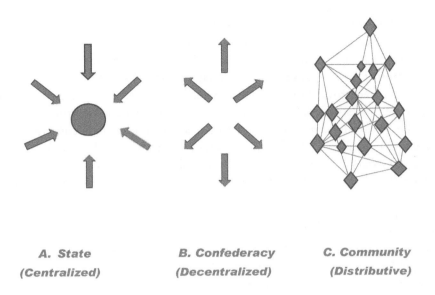

A. State	B. Confederacy	C. Community
(Centralized)	(Decentralized)	(Distributive)

Figure 4.1: Three Governance Models, schematically shows the lines of authority in statist, confederated, and community relationships.
Source: Author.

Once a sufficiently diffuse set of colonies are seized to monopolize their drainable resources, then the serial re-raiding that results is dubbed "imperialism." Seized lands are now looted at organized will by the state. Empires grow through the mass raiding of warfare, as when the British Empire "won" North America from France in 1763—or, for that matter, as when the newly imperial United States "won" the Old Northwest from England in the American Revolution.[18] Whenever territory has become too dispersed to be patrolled and controlled militarily for the imperial center, the empire collapses. For instance, the United States has fought to maintain its foreign oil empire in the Middle East, even as its periphery closes in on the U.S. center. Throughout the evolution of the raid from piracy to empire, there is only one line of authority, regardless of whether the governmental structure is apparently democratic or blatantly dictatorial.

Because the raid is obviously a system of diminishing returns when practiced over and over against the same set(s) of resources, scarcity becomes the primary organizing principle of the raid's economic model. Its thrust is to monopolize all the zero sums in sight, because under this model, "more for you is less for me," as Charles Eisenstein so memorably puts the matter.[19] Indigenous famine did not, for instance, set in until the settler raiding economy stole the goods.[20] In fact, the preferred method of handling the "vexed question" of unwanted Indigenous populations became deliberate over-raiding for the purpose of producing Indigenous starvation. For the United States, this tactic started with George Washington's destruction of the Iroquois fields and storehouses in 1779 and continued throughout the nineteenth-century's Western expansion.[21] In an 1875 speech to the Texas legislature in this period, U.S. Army General Philip Sheridan cheerfully predicted that the so-called Indian Problem of America would be overcome in just another three years, because his buffalo hunters were industriously "destroying the Indians' commissary."[22] Sheridan thus laid the raiding principle of scarcity bare of its fancy-dance in the starkness of his (B [Buffalo]) / (I [Indian]) formula, in which: $(1B - 1) = 0I$.

At this point, imperial theorizing begins grandly obscuring the raw raid as "capitalism," whose elaborate flying buttresses spin out into ever more extravagant excrescences. These ideological flourishes fulfill two functions: First, they fool public into thinking that respectability must stand behind all the convoluted props. Second, the elite beneficiaries of "yours is mine" economics can reassure themselves that they are something other than Vikings out on a toot.

By way of contrast, Indigenous matriarchies eschew centralized, hierarchical control. Far from thus demonstrating the "primitive" condition of their societies, the refusal to belong to a state in preference to a community of mutually known, but not mutually chained down, constituents is a conscious and collective choice. Not everyone wants to live in a centralized state. Typically, such dissenters from statism live in matriarchies—and far more happily than they might, should a (neo)-colonial state be imposed on them.[23] This is because, under Indigenous matriarchal systems, power is either decentralized or completely distributive, empowering group consensus over top-down hegemony.

Indigenous decentralization typically takes the form of the great confederacies, for instance, of the northeastern Woodlands Haudenosaunee (Iroquois) culture and the southeastern Woodlands Muscogee culture (Creek).[24] The Shawnee culture is an example of a purely distributive group, its five clans operating independently in space, albeit as connected by gossamer fibers of "family" (clan). Regardless of locale, the Shawnee clans were/are entirely apprised of and cooperative with each other's doings.[25]

For North Americans Indigenes, the two matriarchal forms of confederation and distribution were actually expressions of the Twinned Cosmos, the primary interpretive system of Indigenous America. Under the Sacred Twinship, two interactive and complementary halves exist as covalencies, with each essential to the smooth functioning of the other. Typically, in the Woodlands, the halves are construed as the Blood of Mother Earth, expressed as ♦ (clans), and the Breath of Brother Sky, expressed as »→ (spatiality). West of the Mississippi, the halves are likely to be spoken of as Water (Blood) and Air (Breath), but the same, ♦/»→ dynamic is being referenced.[26] Together, the halves form the balanced, collaborative unit of the whole. Because everything was and still is seen as existing by continually replicating fractals, even the clans and "nations," themselves, exist/ed by halves. In anthro-speak, for Blood ♦ clans the halves are "*moitié*," which is simply French for "half," so that, for instance, the clans of the Iroquois align under either Wolf/(Breath) »→ or Turtle/(Blood) ♦.[27] In terms of the west-called "nations," or the Brotherhoods »→, the Iroquois parse them out as Youngers ♦ (Oneida and Cayuga, including late-coming Tuscaroras) and Elders »→ (Seneca, Onondaga, and Mohawk). The Younger/Elder Twinship is fairly common in "national" halving. Economically speaking, the women maintained the Blood half ♦, which for food production meant fields, whereas the men maintained the Breath half »→, which for food production meant the forests. Thus, the women ran large-scale agricultural concerns, at the same time that the men ran very large, free-range animal preserves.[28] Combined ♦ + »→, the Twinship fed all the people, and quite handsomely. Plenty, not scarcity, is and was the organizing principle.[29]

The great Woodland confederacies were actually complexes of distributive groups. As such, they were far more sophisticated than has been yet been appreciated by any Western scholar I have read. Each of the spatial structures »→, commonly called "nations" by the setters, were typically composed of male hunting societies as underpinned by the whole of the distributive communities ♦, composed of the matrilineal clans. Thus, the Woodland Confederacies only looked like European-style, territorially based nations to the Europeans who, speaking only to the men, completely missed the female fact that the glue holding the confederated spaces together were the several "inside" Blood clans operating on the distributive principle.

The clans' representatives were strewn throughout the "outside" Breath collectives, as shown in Figure 4.2, below. The Iroquois conceive of these interactive halves as the "Father" »→ and "Mother"♦ sides of the League.[30] ("Father" is a stylized reference to Breath, which includes spatiality, while "Mother" is a stylized reference to Blood, which includes the clans.) Breath »→ spatiality is completely interpenetrated by Blood ♦ clans, whose flagella-like bonds lash them together in a spider-webbing of ties. No one cilium (matrilineage, of which there may be several in any one clan) is strong, in and of itself, but their repeating linkages bundle the whole into very tough bindings. Given such imagistic thinking, it is hardly accidental that Indigenous Americans excelled at textile work, twining fiber into cables that were stronger than Spanish steel.[31]

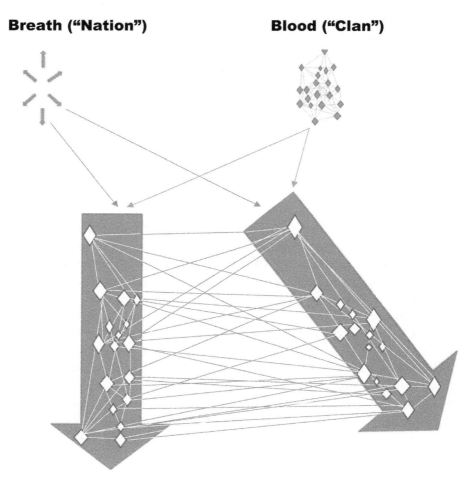

Figure 4.2: Shows the complex intertwinings of Blood ("clans") and Breath ("nations") binding the Confederacies into strong units.
Source: Author.

This concept of intertwinings of tiny strength is precisely that, using which, the Haudenosaunee Peacemaker performed the theory of the Iroquois League for the assembled people in the year 1142.[32] Taking one arrow, standing for just one distributive complex »→/◆, he broke it easily. Then, taking five arrows, standing for all five, interlocking distributive complexes of the original League, 5(»→/◆), he showed that, as a bundled unit, they could not be broken by the strongest force.[33] The League's structure indeed proved so strong and lasting, and so good at keeping the peace, that by the time of European invasion, over sixty Woodlands groups were affiliated with it.[34]

The Blood ◆ clans bind the Breath »→ spatialities together through the gift. The first giver is Mother Earth, as all Indigenous cultures of North America have always maintained. Woodlands crops are planted in what Europeans called "planting mounds," but these conical mounds are traditionally viewed as the Breasts of Mother Earth, upon which her children suckle.[35] She nourishes her children at these breasts, not as an act of scarcity-based rationing, but as an act of plenty-based renewability. In the Plenty Way, all the children are equally fed, with no thought of demanding anything in return, much less of taking advantage of the child through an "exchange," in which the mother greedily and deceptively grasps for more than she has given, as in the European raiding system of capitalism.[36] Instead, in a mirroring of the mother's generosity, the child takes only what s/he needs.

Plenty renews endlessly in this model, using the Indigenous perception of circular return, seen in the halved sharing of the sky, for instance, by the moon and the sun, or in the halved or double-halved sharing of the seasons, being summer »→/winter ◆. (In places that have four seasons, they parse out as summer »→ fall to winter ◆ spring.) In a standing Woodlands idiom, when a male impregnates a female, it is said that he "breathed" on her. The imagery behind this idiom came from observing another of the many natural performances of endlessly renewing plenty, for after the fall crops ◆ are harvested, Father Air *breathes* »→ on Mother Earth, fertilizing Her spring, anew. Just as the man never runs out of semen, the air never runs out of wind.

Both matriarchy and its gift economy must be present for the Indigenous system to sustain itself. One half divorced from its Sacred Twin collapses the structure, as we clearly see in the current American system, as it dwindles down, today. Despite having appropriated the Indigenous, tripartite balance of power among executive, deliberative, and judicial branches of the Tree of Peace, the U.S. Constitution left out the leavening of a twinning system.[37] Indigenously, it is not just men who meet in council, but women, who meet in parallel councils. Only interactively between the halves, does the Twinship function.[38] Indigenously, Grandmothers' as well as Grandfathers' councils—with the balance tipped to the Grandmothers[39]—ran the full-on gift economy, in which all goods were delivered to, retained, and distributed by the Clan Mothers.[40] In drawing up its society from the Iroquoian model, the settler government just ignored the women, while instating mercantile capitalism in its copycat system, leaving settler women to fight for equality while instating what amounted to the raiding economics of Old Europe.[41] For all the words spewed out of academia regarding the Iroquois League, these most basic facts of its operation have been studiously obviated, but then, missing the Indigenous point is a commonplace and, some of us think, a goal of neo-colonial scholarship.

Under the imagery of Vaughan's model, the parasite of capitalism has certainly been effective in consuming its prey, the Indigenous gift economy, over the last two thousand years.[42] However, the raid's essential premise of "rip and run" is now down to eating through its own internal organs. There is simply nothing left to raid. All the Earth is robbed of Her resources, fouled by the wastes of heedless raiding, so that climate change in the present is just the beginning of a massive, modern species extinction event.[43] Oceans suffer dead zones even as they flood city-lined shorelands, while the air becomes unbreathable, "fresh" waters become unpotable, and the poles migrate, wobbling amiss, at the precise moment that astronomers realize with an unpleasant jolt just how many "near-earth objects" there really are buzzing our beleaguered planet.[44]

The science of environmental collapse is indisputable, yet "The Raid" remains stuck in the fifth-century mentality fostered by ignorant and violent elites who plan to survive any changes unscathed, cushioned by their wealth and privilege, as always. In their minds, their statism needs, therefore, no fallback position but to find new fields of plunder, hence the eager push today to migrate into outer space, with moon colonies, Mars colonies, and "exploration" seeking new Earths.[45] In a holding pattern, till something turns up, capitalist states circle past the locus of alarm, officially denying that the climate is folding, quietly downplaying the melting polar sheets, and confining to an isolated "science" community news of the new axis tilt of the migrating poles and the earth's cosmic close calls. All the while, the vast and dumbed-down public is urgently prompted to *look over there! à la folie du jour* of one state-manufactured, distracting "crisis" after another.[46] This is the oft-repeated course of Western empires, but it is certainly not the only course open to the world. There is also the shining matriarchal model, which offers a stable and peaceful alternative in its gift economy.

Discussion Questions

Why are "Western" formulations still allowed to guide all intellectual discussions?

Hawaiian scholar, Haunani-Kay Trask, describes a main element of raiding culture economics as "predatory individualism" (Trask, 4, 5).[47] How do you see the Western concept of individualism applying to this conversation?

Is the environmental destruction of Earth (or the Moon, or Mars, etc.) necessarily an outcome of current space programs?

Indigenous Americans see a multiplicity of ties through "fictive kinship" as an important element of distributive economics. How does this compare with the mechanisms of a capitalist marketplace?

Notes

1. Barbara Alice Mann, "The Raid," in *Verantwortung—Anteilnahme—Dissidenz: Patriarchatskritik als Verteidigung des Lebendigen*, eds. Mathias Behmann, Theresa Frick, Ursula Scheiber, and Simone Wörer (Frankfurt: Peter Lang, 2013), 79–85.
2. Bruce E. Johansen, ed., *The Encyclopedia of Native American Economic History* (Westport, CT: Greenwood Press, 1999), xiv, 112–13, 129–34; Barbara Alice Mann, *Iroquoian Women: The Gantowisas* (New York, NY: Peter Lang, 2000), 230–37.
3. Hippocrates, *The Genuine Works of Hippocrates in Two Volumes*, vol. 1 (New York, NY: William Woodand Company, 1886), 174; Carle Clark Zimmerman and Lucius Ferdinand Cervantes, *Marriage and Family: A Text for Moderns* (Washington, D.C.: Henry Regnery Company, 1956), 297.
4. Denis Delâge, *Le pays renversé: Amérindiens et Européens en Amérique du Nord-Est, 1600–1664* (Montréal: Boréal Express, 1985), 64–65.
5. Adam Arvidsson and Nicolai Peiterson, *The Ethical Economy: Rebuilding Value after the Crisis* (New York, NY: Columbia University Press, 2013), 166–67 n. 2; Eric S. Raymond, *The Cathedral and the Bazaar: Musings on Linux and Open Source by an Accidental Revolutionary*, revised & expanded ed. (Sebastopol, CA: O'Reilly Media, 2001), 84–86.
6. Lewis Henry Morgan, *Ancient Society: Researches in the Lines of Human Progress from Savagery through Barbarism to Civilization* (New York, NY: Henry Holt and Company, 1877), 17.
7. Simone de Beauvoir, *The Second Sex*, new translation by Constance Borde and Sheila Malovany-Chevallier (New York: Alfred Knopf, 2010), 63, 568.
8. Ibid., 72.
9. Betty Friedan, *The Feminine Mystique* (New York, NY: W. W. Norton, 1963), 253, 348. For childrearing as "drudgery," 18; for and "drudgery" characterizing childcare, see quotation, 201.
10. Cynthia Ellers, *The Myth of Matriarchal Prehistory: Why an Invented Past Won't Give Women a Future* (Boston, MA: Beacon Press, 2000).
11. Nancy Dowd and Michelle S. Jacobs, *Feminist Legal Theory: An Anti-Essentialist Reader* (New York, NY: New York University Press, 2003), 6; Cheikh Anta Diop, *Civilization or Barbarism: An Authentic Anthropology* (New York, NY: Laurence Hill Books, 1991), 121–22; Ifi Amadiume, *Reinventing Africa: Matriarchy, Religion, Culture* (New York, NY: St. Martin's Press, 1997), for "juridical" as descriptor, 78.
12. Barbara Alice Mann and Heide Goettner-Abendroth, *Matriarchal Studies: A Bibliography*, Oxford University Press Bibliographies (New York, NY: Oxford University Press, 2015), http://www.oxfordbibliographies.com/view/document/obo-9780199766567/obo-9780199766567-0113.xml?rskey=JvD9DR&result=1&q=matriarchal+studies#firstMatch;

Heide Goettner-Abendroth, *Matriarchal Societies: Studies on Indigenous Cultures across the Globe* (New York, NY: Peter Lang, 2012).

13. Genevieve Vaughan, *For-giving: A Feminist Criticism of Exchange* (Austin, TX: Plain View Press, 1997). For her use of "*homo donans*," 46; for "*homo economicus*," 52–53.

14. Steven T. Newcomb, *Pagans in the Promised Land: Decoding the Doctrine of Christian Discovery* (Golden, CO: Fulcrum, 2008), see esp. 59–60, 103–104; Vaughan, *For-giving*, 181.

15. Vaughan, *For-giving*, 42–47, 61.

16. George S. Tinker, *American Indian Liberation: A Theology of Sovereignty* (Maryknoll, NY: Orbis Books, 2008), 8, 52.

17. For an unvarnished look at the "founding" of the British West Indies as pure acts of piracy, see Richard S. Dunn, *Sugar and Slaves: The Rise of the Planter Class in the English West Indies, 1624–1713* (Chapel Hill, NC: University of North Carolina Press for the Institute of Early American History and Culture, 1972). For material on Drake, specifically, see 10–11, 20; for all in the period, see 3–45.

18. Alfred A. Cave, *The French and Indian War* (Westport, CT: Greenwood, 2004); Barbara Alice Mann, "The Greenville Treaty of 1795: Pen-and-Ink Witchcraft in the Struggle for the Old Northwest," in *Enduring Legacies: Native American Treaties and Contemporary Controversies*, ed. Bruce E. Johansen (Westport, CT: Praeger Press, 2004), 135–202.

19. Charles Eisenstein, *Sacred Economics: Money, Gift, and Society in the Age of Transition* (Berkeley, CA: Evolver Editions, 2011), 9, 121–24.

20. See, for instance, the effect on the High Plains peoples of the organized crime called the "fur trade," discussed in Barbara Alice Mann, *The Tainted Gift: The Disease Method of Frontier Expansion* (Santa Barbara, CA: ABC-CLIO, 2009), 46–47.

21. Barbara Alice Mann, *George Washington's War on Native America* (Westport, CT: Praeger, 2005), 27–110.

22. Quotations from Sheridan's 1875 speech, recorded in John R. Cook, *The Border and the Buffalo: An Untold Story of the Southwest Plains, A Story of Mountain and Plain* (Topeka, KS: Crane and Company, 1907), 113.

23. Amadiume, *Reinventing Africa*, 139. See Amadiume's entire discussion and example of the imposition of a state on a non-state community, 109–40.

24. Multiple works examine the functioning of the Iroquoian Confederacy, modernly including Mann, *Iroquoian Women*. The classic works, including Lewis Henry Morgan, *League of the Haudenosaunee, or Iroquois*, 2 vols. (1851; reprint, New York, NY: Burt Franklin, 1901) must be read judiciously, however, for they are awash in Westernizations and misunderstandings. Some modern histories of the Muscogee which indicate confederate operations, albeit without specifically examining the confederacy, *per se*, include Robbie Ethridge, *Creek Country: The Creek Indians and Their World* (Chapel Hill, NC: The University of North Carolina Press, 2003) and Claudio Saunt, *A New Order of Things: Property, Power, and the Transformation of the Creek Indians, 1833–1816* (Cambridge, UK: Cambridge University Press, 1999). The "classic" source on the Muscogee Confederacy is, unfortunately, also old and skewed, that being Benjamin Hawkins, *Creek Confederacy and a Sketch of the Creek Country* (1848; reprint London: Forgotten Books, 2015).

25. Erminie Wheeler Voegelin, *Mortuary Customs of the Shawnee and other Eastern Tribes*, Prehistory Research Series, vol. 2, no. 4, 1944 (Reprint; New York, NY: AMS Press, 1980), 243, 256.

26. For a complete discussion of the operation of this principle, see Barbara Alice Mann, *Spirits of Blood, Spirits of Breath: The Twinned Cosmos of Indigenous America* (New York, NY: Oxford University Press, 2016).

27. Mann, *Iroquoian Women*, 93–97, and fig. 2.2.; for fractalization, see Barbara Alice Mann, "All My Relatives: The Binary Fractals of the Gift Economy," in *What Comes after Money?: Essays from Reality Sandwich on Transforming Currency and Community*, ed. Daniel Pinchbeck and Ken Jordan (Berkeley, CA: Evolver Editions, 2011), 58–66.

28. Mann, *Iroquoian Women*, for field and farming, see pp. 197, 214–29; for forest preserves, 191–93. This schema was commonly used throughout the Woodlands.

29. Ibid, 202–203.

30. John Napoleon Brinton Hewitt, "Some Esoteric Aspects of the League of the Iroquois," *Proceedings of the International Congress of Americanists* 19 (1915): 325; John Napoleon Brinton Hewitt, "Ethnological Studies among the Iroquois Indians," *Smithsonian Miscellaneous Collections* 78 (1927): 240–41.

31. Charles C. Mann, *1491: New Revelations of the Americas before Columbus*, 2nd ed. (New York: Vintage Books: 2011), 95–96.

32. As Fields and I showed in 1997, the League was founded in 1142, Barbara Alice Mann and Jerry L. Fields, "A Sign in the Sky: Dating the League of the Haudenosaunee," *American Indian Culture and Research Journal* 21, no. 2 (1997): 105–63.

33. Arthur Caswell Parker, "The Constitution of the Five Nations," *New York State Museum Bulletin* 184 (April, 1916): 45; United States Senate, Committee on Indian Affairs, *Native American Free Exercise of Religion Act: Hearing Before the Committee on Indian Affairs*, 103rd Congress, 1st Session, on S. 1021, "To Assure Religious Freedom to Native Americans" (Washington, D.C.: Government Printing Office, 1993), 89.

34. John Reed Swanton, "Iroquois" (list of affiliations), *The Indian Tribes of North America* (Washington, D.C.: Government Printing Office, 1952), 33–39.

35. Arthur Caswell Parker, *Iroquois Uses of Maize and Other Plants* (Albany, NY: University of the State of New York, 1910), 36–37.

36. I first proposed the "Plenty Way" in 2000, in Mann, *Iroquoian Women*, 204, 211.

37. For copying the Iroquois structure, see the classics, Bruce E. Johansen, *Forgotten Founders: Benjamin Franklin, the Iroquois and the Rationale for the American Revolution* (Ipswich, MA: Gambit Incorporated, Publishers, 1982); and Donald A. Grinde, Jr., and Bruce E. Johansen, *Exemplar of Liberty* (Los Angeles, CA: University of California at Los Angeles, American Indian Studies Center, 1991). For my take on the structure and the gift economy of the League, see Mann, *Iroquoian Women*, 161–64, for capitalism, 202–205, for gift economy, 228–37. For the Iroquonian gift economy, see Goettner-Abendroth, *Matriarchal Societies*, 312–14.

38. Bruce E. Johansen and Barbara Alice Mann, eds., *The Encyclopedia of the Haudenosaunee (Iroquois Confederacy)* (Westport, CT: Greenwood Press, 2000), for "Governmental Functioning and Powers of the Haudenosaunee League," see 122–31.

39. The Iroquois Constitution goes out of its way to say that the men had the same rights as women, Parker, *Constitution*, 55.
40. For women's receipt and control of the food supply, see, for instance, Mann, *Iroquoian Women*, 107–108, 228, 254, 286.
41. For the U.S. Constitution as derived heavily from the Iroquois Constitution, see Johansen, *Forgotten Founders*, and Grinde and Johansen, *Exemplar of Liberty*.
42. Vaughan, *For-giving*, 121, 207, 345.
43. Gerardo Ceballos, et al., "Accelerated Modern Human–induced Species Losses: Entering the Sixth Mass Extinction," *Science Advances* 1, no. 5 (19 June 2015), accessed 15 April 2016, http://advances.sciencemag.org/content/advances/1/5/e1400253.full.pdf.
44. Cheryl Lyn Dybas, "Dead Zones Spreading in World Oceans," *BioScience* 55, no. 7 (2005): 552–57; A. Dutton, et al., "Sea-level Rise due to Polar Ice-sheet Mass Loss during Past Warm Periods," *Science*, 349, no. 6244 (10 June 2015): 153 *et seq.*, accessed 15 April 2016, doi: 10.1126/science.aaa4019; Nils Olsen and Miaora Mandea, "Will the Magnetic North Pole Move to Siberia?" *Eos* 88, no. 29 (2007): 293–300; Cameron McPherson Smith and Evan Tyler Suliëman Davies, *Emigrating beyond Earth: Human Adaptation and Space Colonization* (New York, NY: Springer, 2012), 123–26.
45. Smith and Davies, *Emigrating beyond Earth*, 116–17.
46. See the classic on the art of manipulating crisis and accord for political gain, Edward S. Herman and Noam Chomsky, *Manufactured Consent: The Political Economy of the Mass Media*, rev. ed. (1988; New York: Pantheon Books, 2002), for an example of conveniently distracting crisis, see 137–39.
47. Haunani-Kay Trask, *From a Native Daughter: Colonialism and Sovereignty in Hawai'I*, rev. ed. (Honolulu: University of Hawai'i Press, 1999) 4, 5.

References

Amadiume, Ifi. *Reinventing Africa: Matriarchy, Religion, Culture.* New York, NY: St. Martin's Press, 1997.

Arvidsson, Adam, and Nicolai Peiterson. *The Ethical Economy: Rebuilding Value after the Crisis.* New York, NY: Columbia University Press, 2013.

Beauvoir, Simone de. *The Second Sex.* New translation by Constance Borde and Sheila Malovany-Chevallier. New York: Alfred Knopf, 2010.

Cave, Alfred A. *The French and Indian War.* Westport, CT: Greenwood, 2004.

Ceballos, Gerardo, Paul R. Ehrlich, Anthony D. Barnosky, Andrés García, Robert M. Pringle, and Todd M. Palmer. "Accelerated Modern Human–Induced Species Losses: Entering the Sixth Mass Extinction." *Science Advances* 1, no. 5 (19 June 2015), accessed 15 April 2016. http://advances.sciencemag.org/content/advances/1/5/e1400253.full.pdf.

Cook, John R. *The Border and the Buffalo: An Untold Story of the Southwest Plains, A Story of Mountain and Plain.* Topeka, KS: Crane and Company, 1907.

Delâge, Denis. *Le pays renversé: Amérindiens et Européens en Amérique du Nord-Est, 1600–1664.* Montréal: Boréal Express, 1985.

Diop, Cheikh Anta. *Civilization or Barbarism: An Authentic Anthropology.* New York, NY: Laurence Hill Books, 1991.

Dowd, Nancy, and Michelle S. Jacobs. *Feminist Legal Theory: An Anti-Essentialist Reader.* New York, NY: New York University Press, 2003.

Dunn, Richard S. *Sugar and Slaves: The Rise of the Planter Class in the English West Indies, 1624–1713.* Chapel Hill, NC: University of North Carolina Press for the Institute of Early American History and Culture, 1972.

Dutton, A., A. E. Carlson, A. J. Long, G. A. Milne, P. U. Clark, R. DeConto, B. P. Horton, S. Rahmstorf, and M. E. Raymo. "Sea-level Rise due to Polar Ice-sheet Mass Loss during Past Warm Periods." *Science* 349, no. 6244 (10 July 2015): 153. Accessed 15 April 2016. doi: 10.1126/science.aaa4019.

Dybas, Cheryl Lyn. "Dead Zones Spreading in World Oceans." *BioScience* 55, no. 7 (2005): 552–57.

Eisenstein, Charles. *Sacred Economics: Money, Gift, and Society in the Age of Transition.* Berkeley, CA: Evolver Editions, 2011.

Ellers, Cynthia. *The Myth of Matriarchal Prehistory: Why an Invented Past Won't Give Women a Future.* Boston, MA: Beacon Press, 2000.

Ethridge, Robbie. *Creek Country: The Creek Indians and Their World.* Chapel Hill, NC: The University of North Carolina Press, 2003.

Friedan, Betty. *The Feminine Mystique.* New York, NY: W. W. Norton, 1963.

Goettner-Abendroth, Heide. *Matriarchal Societies: Studies on Indigenous Cultures across the Globe.* New York, NY: Peter Lang, 2012.

Grinde, Donald A., Jr., and Bruce E. Johansen. *Exemplar of Liberty.* Los Angeles, CA: University of California at Los Angeles, American Indian Studies Center, 1991.

Hawkins, Benjamin. *Creek Confederacy and a Sketch of the Creek Country.* 1848. Reprint, London: Forgotten Books, 2015.

Hippocrates. *The Genuine Works of Hippocrates in Two Volumes*, vol. 1. New York, NY: William Wood and Company, 1886.

Herman, Edward S., and Noam Chomsky. *Manufactured Consent: The Political Economy of the Mass Media.* Rev. ed. New York, NY: Pantheon Books, 2002.

Hewitt, John Napoleon Brinton. "Ethnological Studies among the Iroquois Indians." *Smithsonian Miscellaneous Collections* 78 (1927): 237–47.

———. "Some Esoteric Aspects of the League of the Iroquois." *Proceedings of the International Congress of Americanists* 19 (1915): 322–26.

Johansen, Bruce E. *Forgotten Founders: Benjamin Franklin, the Iroquois and the Rationale for the American Revolution.* Ipswich, MA: Gambit Incorporated, Publishers, 1982.

———., ed. *The Encyclopedia of Native American Economic History.* Westport, CT: Greenwood Press, 1999.

Johansen, Bruce E., and Barbara Alice Mann, eds. *The Encyclopedia of the Haudenosaunee (Iroquois Confederacy).* Westport, CT: Greenwood Press, 2000.

Mann, Barbara Alice. "All My Relatives: The Binary Fractals of the Gift Economy." In *What Comes after Money? Essays from Reality Sandwich on Transforming Currency and Community*, edited by Daniel Pinchbeck and Ken Jordan, 58–66. Berkeley, CA: Evolver Editions, 2011.

————. *George Washington's War on Native America*. Westport, CT: Praeger, 2005.

————. "The Greenville Treaty of 1795: Pen-and-Ink Witchcraft in the Struggle for the Old Northwest." In *Enduring Legacies: Native American Treaties and Contemporary Controversies*, edited by Bruce E. Johansen, 135–202. Westport, CT: Praeger Press, 2004.

————. *Iroquoian Women: The Gantowisas*. New York, NY: Peter Lang, 2000.

————. "The Raid." In *Verantwortung—Anteilnahme—Dissidenz: Patriarchatskritik als Verteidigung des Lebendigen*, edited by Mathias Behmann, Theresa Frick, Ursula Scheiber, and Simone Wörer, 79–85. Frankfurt: Peter Lang, 2013.

————. *Spirits of Blood, Spirits of Breath: The Twinned Cosmos of Indigenous America*. New York, NY: Oxford University Press, 2016.

————. *The Tainted Gift: The Disease Method of Frontier Expansion*. Santa Barbara, CA: ABC-CLIO, 2009.

Mann, Barbara Alice, and Jerry L. Fields. "A Sign in the Sky: Dating the League of the Haudenosaunee." *American Indian Culture and Research Journal* 21, no. 2 (1997): 105–63.

Mann, Barbara Alice, and Heide Goettner-Abendroth. *Matriarchal Studies: A Bibliography*. Oxford University Press Bibliographies. New York, NY: Oxford University Press, 2015. http://www.oxfordbibliographies.com/view/document/obo-9780199766567/obo-9780199766567-0113.xml?rskey=JvD9DR&result=1&q=matriarchal+studies#firstMatch.

Mann, Charles C. *1491: New Revelations of the Americas before Columbus* (2nd ed.). New York, NY: Vintage Books, 2011.

Morgan, Lewis Henry. *Ancient Society: Researches in the Lines of Human Progress from Savagery through Barbarism to Civilization*. New York, NY: Henry Holt and Company, 1877.

————. *League of the Haudenosaunee, or Iroquois* (2 vols.). 1851. Reprint, New York, NY: Burt Franklin, 1901.

Newcomb, Steven T. *Pagans in the Promised Land: Decoding the Doctrine of Christian Discovery*. Golden, CO: Fulcrum, 2008.

Olsen, Nils, and Miaora Mandea. "Will the Magnetic North Pole Move to Siberia?" *Eos* 88 no. 29 (2007): 293–300.

Parker, Arthur Caswell. "The Constitution of the Five Nations." *New York State Museum Bulletin* 184 (April, 1916): 7–158.

————. *Iroquois Uses of Maize and Other Plants*. Albany, NY: University of the State of New York, 1910.

Raymond, Eric S. *The Cathedral and the Bazaar: Musings on Linux and Open Source by an Accidental Revolutionary*. Revised and expanded ed. Sebastopol, CA: O'Reilly Media, 2001.

Saunt, Claudio. *A New Order of Things: Property, Power, and the Transformation of the Creek Indians, 1833–1816*. Cambridge, UK: Cambridge University Press, 1999.

Smith, Cameron McPherson, and Evan Tyler Suliëman Davies. *Emigrating beyond Earth: Human Adaptation and Space Colonization*. New York, NY: Springer, 2012.

Swanton, John Reed. "Iroquois." *The Indian Tribes of North America*. Washington, D.C.: Government Printing Office, 1952. 33–39.

Tinker, George S. *American Indian Liberation: A Theology of Sovereignty*. Maryknoll, NY: Orbis Books, 2008.

Trask, Haunani-Kay. *From a Native Daughter: Colonialism and Sovereignty in Hawai'i.* Rev. ed. Honolulu: University of Hawai'i Press, 1999.

United States. Senate. Committee on Indian Affairs. *Native American Free Exercise of Religion Act: Hearing Before the Committee on Indian Affairs.* 103rd Congress, 1st Session. On S. 1021, "To Assure Religious Freedom to Native Americans." Washington, DC.: Government Printing Office, 1993.

Vaughan, Genevieve. *For-giving: A Feminist Critique of Exchange.* Austin, TX: Plain View Press, 1997.

Voegelin, Erminie Wheeler. *Mortuary Customs of the Shawnee and other Eastern Tribes.* Prehistory Research Series, vol. 2, no. 4. 1944. Reprint, New York, NY: AMS Press, 1980.

Zimmerman, Carle Clark, and Lucius Ferdinand Cervantes. *Marriage and Family: A Text for Moderns.* Washington, D.C.: Henry Regnery Company, 1956.

CHAPTER FIVE

Original Practices for Becoming and Being Human[1]

DARCIA NARVAEZ

One must go outside of written history of civilizations to realize that human behavior seen in Westernized industrialized nations is not representative of humanity generally. We all know vicious people or characters in stories—people who are routinely self-centered, cruel, domineering, or exclusionary—people whose "hearts" seem to be missing, who are soul-less, mind-less, grasping for power or control, or those who withdraw from relational connection and self-medicate with consumerism, physical pleasures, media, drugs, monetary or even intellectual pursuits. Why are there so many people like this in the modern world? Why has industrialized culture promoted these outcomes?

For some time, theorists from Rousseau to Freud, Fromm, Mumford, and Montagu have noted the discontentment and even alienation of industrialized humanity.[2] Freud thought civilization necessary to control the wild id in every individual, with the side effect of unhappiness presumably due to thwarted impulses.[3] Others pointed to the alienation from living or working among strangers. Still others pointed to the way the industrialized system, then capitalism, undermines what makes humans truly happy—close family relationships, enjoyable work and leisure, plus enough resources to make those possible. But most industrialized people seem to assume "there is no other way to be," and that whatever is happening is "the price of progress." We are all quite immersed in this worldview, making it difficult to realize that the view is recent and rare in the history of the human species.[4]

The backdrop for discussions in this chapter and book is the whole of humanity and its nature. The setting is an earth community devastated by a particular worldview and cultural practices that uphold and perpetuate it. According to Redfield, there are two basic worldviews: an ancient one that considers the cosmos unified, sacred and moral, and the other that considers the cosmos fragmented, disenchanted, and amoral.[5] The latter, a particularly damaging worldview, evolved in the Old (Western) World, with seeds in Mesopotamia, only a few thousand years ago with the rise of the Abrahamic religions. Taking a stance against embodied life, they pushed their god out of nature as well as themselves. Historian Frederick W. Turner masterfully points to the cultivated fear of wilderness, of animals and nature, and their connotation with evil and depravity, initiated by the sky-god patriarchies that formed in the Middle East.[6] Their nature-divorced god pulled the human narrative off the earth and into a numinous realm to be fully experienced after death. On the earth, all was *de*sacralized. Infiltrating Europe and encouraged by Biblical texts, soon the elites believed that only humans had value, making all else objects for human use.[7] With time, this shifted to a grand scale of imperial entitlement in Western Europe, which, when combined with Enlightenment science, formed a culture was like no other—one "more materialist ... more expansionist, more volatile and energetic, more linked to growth and progress, and almost everywhere without the kinds of moral inhibitions found in the world's other high cultures."[8] Historian William Woodruff noted:

> No civilization prior to the European had occasion to believe in the systematic material progress of the whole human race; no civilization placed such stress upon the quantity rather than the quality of life; no civilization drove itself so relentlessly to an ever—receding goal; no civilization was so passion-charged to replace what is with what could be; no civilization had striven as the West has done to direct the world according to its will; no civilization has known so few moments of peace and tranquility.[9]

What breeds such passion for domination? In regards to the natural world, Western culture also was unusual:

> Its fundamental regard for nature was more hostile and antagonistic than was true for any other developed civilization ... nowhere else was the essential reverence for nature seriously challenged, nowhere did there emerge the idea that human achievement and material betterment were to be won by *opposing* nature, nowhere any equivalent to that frenzy of defiance and destruction that we find on the Western record. ... the central religions of neither the Asian nor the American civilizations permitted a separation from, or an attitude of dominion over, the natural beings and patterns of the nonhuman world.[10]

Daniel Quinn in his novel, *Ishmael*, makes an insightful summary of principles underlying the treatment of the natural world, comparing what he calls "Taker" culture (settled societies dependent on monoagriculture) and "Leaver" culture (nomadic foragers).[11] The book's protagonist identifies how Taker or civilized culture defies the laws of nature. Unlike every other lifeform, civilized humans not only perceive other creatures as competitors but want to: (1) exterminate competitors, (2) systematically destroy competitors' food to make room for humanity's food, (3) deny competitors access to food, and (4) store food extensively only for themselves. We could add to his list: (5) consider the rest of the natural world to be "resources" reserved only for them, (6) use biocides to kill "pests" (which harm the web of life generally). Clearly, it is these attitudes and behaviors that contribute heavily to the dire straits in which humanity and life on the earth find themselves.

This "spirit against the wilderness" was brought by the Old World invaders to the New World with a grasping for profit or control at any cost and an inability to appreciate what they encountered.[12] Many of the invaders were *malcriados* (ill-bred from a human virtue standpoint) who demonstrated the vicious temperaments apparently admired at the time by their cohorts—arrogant, rapacious, tough, exploitative, and cruel. Roy Robbins pointed out that many who took over the land were speculators rather than settlers who, like the conquistadors before them, would carve up the newly plundered lands and then move on.[13] In the introductory chapter of this book, it was mentioned that the invaders acted to chop down, burn or exterminate the living across the land—whether plant, animal or human. What breeds such restlessness? What breeds such repudiation of the unfamiliar other?

Natives peoples of the Americas showed a very different worldview and human nature.[14] According to Arawak scholar, Frank Bracho, happiness for the Indigenous worldview was a matter of *being* more than of *having*—"being in the form most conducive to happiness was linked to the greatest possible integration with creation, with the natural order and its laws."[15] Old World fears clashed with this earth-centric worldview practiced among native peoples that exhibited a joyous appreciation of life in all its forms and a respect for nature's life-death-life cycle which included humans themselves. The nature of the New World natives was remarked upon by explorers and first contact diarists, with a common assessment of their nature. Columbus wrote about the natives he first encountered to his sovereigns, Isabella and Ferdinand, about 10 weeks into his first visit:

> They are so affectionate and have so little greed and are in all ways so amenable that I assure your Highnesses that there is in my opinion no better people and no better land in the world. They love their neighbors as themselves and their way of speaking is the sweetest in the world, always gentle and smiling.[16]

Columbus' gracious attitude was not accompanied by gracious behavior. Columbus nearly immediately started to kidnap and enslave natives. Bartolomé de Las Casas, churchman then priest then landowner, was eyewitness to the developments in Caribbean islands from 1500 on, watching the arrival of "criminals and hoodlums" to populate the place, the enslavement and rape of the natives, the importation of Africans as slaves to replace the dwindling population of natives.[17] After about 15 years, Las Casas had second thoughts about the whole enterprise, reversing his earlier belief that the New World provided an opportunity for Christian evangelism. He began instead to express admiration for the Indian cultures, their rightness for Indian souls, unlike the imported tortures of the Christian invaders.[18]

Native peoples were astonished by the European invaders who behaved viciously, unlike any human beings they had known. For example, Acuera, the Timucua chief, had this to say to the Spanish explorer Hernando de Soto around 1540:

> Others of your accursed race have, in years past, poisoned our peaceful shores. They have taught me what you are. What is your employment? To wander about like vagabonds from land to land, to rob the poor, to betray the confiding, to murder in cold blood the defenceless.[19]

As another read into the natures of the invaders and the natives, note what a philanthropist "friend of the Indians" said of the native in the 19th century:

> *We need to awaken in him wants.* In his dull savagery he must be touched by the wings of the divine angel of discontent ... Discontent with the tepee and the starving rations of the Indian camp in winter is needed to get the Indian out of the blanket and into trousers—and trouser with a pocket in them, and with a *pocket that aches to be filled with dollars!*[20]

In other words, unfettered greed was considered normal, and the natives had to be taught to be greedy.

Whence the origins of these disparate worldviews? Paul Shepard argued that Western/civilized culture arrested development in adolescence due to the lack of challenging initiation rites that bind the individual to the cosmos.[21] I agree that this may be a significant part of the problem but I think that the source of arrested development starts much earlier, in early childhood. In industrialized nations, this is when a deep sense of insecurity towards self, others and the world is fostered, and the child carries throughout life dysregulation and illfittedness with human and other-than-human worlds (unless later significant events alter established patterns), leading to the many psychopathologies that psychoanalysts and others have documented.[22] In contrast, among the many practices that were denigrated and dismissed by European invaders to the Americas was the tender way that children were raised, documented recently among small-band hunter-gatherer societies (nomadic foragers) around the world.[23]

My contention here is that the differences in worldview and human nature begin with early life experiences that shape the brain in a species-typical or atypical manner. What caused civilizations to move away from meeting children's needs is a complicated story I do not address here. However, what is important to note are the differences between societies that do and do not provide for children's full basic needs. Those who provide for children's needs raise adults who live in relative harmony and happiness. Let's see why that is the case.

The Importance of Early Nurturing

Ethnographies of societies that live as nomadic foragers—the type of society in which the human genus spent 99% of its history and was common to many native American tribal groups—show us a different view of humanity than do civilized accounts. In these societies, children are affectionately raised by the whole community. Infants form relationships with multiple members of the community. For example, in observational studies they are held by multiple caregivers within the two-hour windows of time observed and recorded; then, past 18 months, they spend time with mother only about 40% of the time.[24] Native Americans also demonstrate the community raising of children. John Mohawk pointed out:

> the group was responsible for life and death. From the moment you were born, they were responsible for you. Not your mother, but the whole group. From the moment you're born, they're all paying attention; it's usually women who are paying attention to you when you're very small. As you get older, especially if you're a male, then the males take over at some point.[25]

In fact, one of the most important evolutionary inheritances humans have may be the cooperative raising of children.[26] This forms part of developmental systems that humans, similar to all animals, evolved to optimize the normal development of their offspring. Nurturing is a large part of our mammalian heritage. Humans are part of the tree of life, sharing many biological characteristics with other animals and inheriting many things from their ancestors beyond genes, including aspects of the human body that resemble those of fish, plants, mammals.

Humans are born highly immature compared to other hominids (and should be in the womb at least another 18 months!)[27] As a result, most brain development occurs after birth. Because we are born so immature, with only 25% of adult brain size, much of our brain and body functions are established after birth through a combination of epigenetics and plasticity interactions with the child's maturational schedule. A child's relational world starts with mother's body, attention, and care. The child's proper development depends on maternal bonding and continued entrainment with her physiology in the first months and continued responsive care in the first years. An immature infant learns neurobiological self-regulation and grows social capacities from calming, affectionate care. Thus, humans evolved to expect particular type of care, an intense level of support on the part of the mother and community, a situation that was available throughout most of humanity's existence.[28]

As ethological observation has noted, all animals provide a nest that matches up with the maturational schedule of their young in order to optimize normal development.[29] Humans are no different. Humans evolved a particular nest to provide the intensive care a newborn needs.[30] Nurturing begins with the *evolved nest*.

How do we know what humanity's evolved nest looks like? Substantive evidence comes from extant studies of foraging communities around the world, the type of society in which the human genus spent 99% of its history.[31] Foragers raise their children in a similar way wherever they have been observed around the world.[32] Anthropologists summarize the shared caregiving for infants and young children across these groups:

> young children in foraging cultures are nursed frequently; held, touched, or kept near others almost constantly; frequently cared for by individuals other than their mothers (fathers and grandmothers, in particular) though seldom by older siblings; experience prompt responses to their fusses and cries; and enjoy multiage play groups in early childhood.[33]

To this list can be added soothing perinatal experiences and the positive social support noted earlier.[34] These nest components are vital for normal development as a human being.

Why does early nurturing matter so much? Because it shapes the functioning of the body/mind. As embodied biosocial "becomings," a person's neurobiology must be well developed. Optimal human functioning depends on how well one's biology functions, like self-regulatory systems, and influences one's sociality, such as empathy and perspective taking. Early nurturing establishes the psychobiosocial nature of the individual that is carried forward, without intervention, throughout life. A well-raised individual shows social fittedness and self regulation "all the way down," displaying both cooperation and calmness.

I briefly note effects of the evolved nest. Mothers are supported and *avoid stress and distress during pregnancy* as these negatively influence the personality and physiology of the child.[35] After normal, *naturalistic birth* conditions, mother does not want to put down baby. She and the baby become magnetized to one another from reward centers being activated in the first hour or so.[36] Babies needs are met without question (*responsivity*). Babies are kept calm as their physiological systems set their parameters and thresholds for functioning in the first couple years of life.[37] *Breastfeeding* take place for years in order to develop the proper physiology of the child, such as the immune system which does not reach adult functioning until around age 5.[38] Babies expect *affection*, to be held and carried virtually all the time—their physiological development requires it.[39] Babies need constant in-arms support, which of course cannot be easily accomplished by a mother alone. In fact, isolating mothers and babies is a recent, health-undermining practice. Mothers in our prehistory were always mothering *with other mothers* who shared care and even breastfeeding when necessary.[40] All mammals (and other animals too) *play*, a sign of health and sense of safety.[41] When young mammals play they build self-control and social skills.[42] Children expect to play from birth.[43] Banter (sometimes called "gossip"), a more verbal form of play, is what characterizes human relationships in our ancestral conditions.[44] Indeed, in the majority of small-band hunter-gatherer communities, most of life is spent *communally*, in social leisure (minimal work is done for food-gathering) which includes song, dance, storytelling and teasing. The evolved nest meets mammalian social-emotional needs, structuring body systems in the first years of life, whose effects last through the lifespan.

One of the most important growth points in early life is the right hemisphere (RH), the half of the brain that governs self-regulation and implicit social relations.[45] This has deep implications for worldview. The Indigenous worldview understands that spirit pervades all things, that all are related and that humans have much to learn from the elders in the world—the plants and animals. These are capacities supported by the RH, capacities that are shaped from early life experience, when the RH is scheduled to grow more rapidly. The RH governs self-regulatory processes as they develop in early life, critical for our awareness of the dynamic interaction of all living things and the uniqueness of lifeforms we encounter,[46] a "process" view of life that fits with theories that are still largely marginalized in Western thought.[47]

Under species-typical conditions, children extend their relational bonding from mother to two more groundings early on. First, the child finds and builds relationships with members of the community of older members through attraction and responsive interaction. These are the *allomothers* that support the child's flourishing. Second, the child becomes rooted in the other-than-human community, which at first offers the backdrop for human relationships but then becomes a deeper set of relationships with animals and plants through attraction and inter-action. Paul Shepard wrote:

> Animals have a magnetic affinity for the child, for each in its way seems to embody some impulse, reaction, or movement that is "like me." In the playful, controlled enactment of them comes a gradual mastery of the personal inner zoology of fear, joy, and relationships. In stories told, their form spring to life in the mind, re-presented in consciousness, training the capacity to imagine.[48]

In hunter-gatherer societies, adolescents move into a lifelong study of recipro-cal relationships with the other-than-human world. Their lives depend on stay-ing connected and fostering the wellbeing of the local biocommunity. Civilized nations are more distant from the natural world and have treated it like a cheap resource.[49] But their wellbeing also requires the wellbeing of the biocommunity.

The Nested Community

Ninety-nine percent of humanity's existence has been spent in nomadic forag-ing communities. In extant studied communities and in reports by first contacts, they held a common worldview, that all things are alive with agency and purpose. There was a common adult personality—generosity happiness, calmness, and high intelligence.[50] Indeed, it was not only infants and children who benefitted from the evolved nest; all age groups were embedded in a nest of support for a lifetime. Anthropologists have noted the intense social fabric of simple hunter-gatherer societies, documenting the prevalence of positive social interactions, such as the frequency of laughter (which has beneficial effects on immune systems), singing (which has improves wellbeing), and constant socializing (which can be related to better health outcomes).[51] Moreover, children and adults in nomadic forag-ing societies experienced purposeful social egalitarianism in the midst of natural dominance patterns—there were no strict culturally imposed social hierarchies.[52] In fact, studies across species, including human children and adults, find nega-tive reactions to perceived unequal treatment.[53] Un-nested environments, those of industrialized nations, may contribute to decreased wellbeing in adolescents and adults as well as in infants and children.

Consequences of the Degraded Nest

When humans, like other animals, don't receive their species-typical evolved nest, they become abnormal specimens—just like Harry Harlow's monkeys, deprived of maternal love from birth, who became autistic and antisocial.[54] Just as a species-typical nest is necessary to foster a species-typical individual, whether monkey or elephant or whale, the evolved nest is fundamental to becoming human. Species-typical animals learn their place in the biocommunity and live cooperatively with other species. Human physiology and psychology are highly shaped by post-natal experience, more so than for any other animal.[55] It may be best illustrated this way. Think about raising a wolf in a human family: you will end up with a wolf. But if you raise a human in a wolf family, you end up with a wolf-child (as has happened numerous times in known history): an individual missing many characteristically human attributes like not walking on their feet but on all fours, and showing wolf-appropriate communications and social skills. Humans are greatly affected by their experiences after birth. Neurobiological systems important for all a human becomes go through sensitive periods to be shaped at this time. This includes the development of sociality and moral values (and Darwin's moral sense).[56]

The United States today especially undermines the provision of what young children need, impairing optimal development at virtually every step of the way. There is *little expectation of calm pregnancies* in the United States these days as all women are expected to work right up to the birth (and even return to work shortly after birth); medicalized birth in the USA is a clear and present danger for normal mothers and at every step tends to traumatize the mother and baby (separation, painful procedures, sensory shock to the infant), undermining normal mother-child bonding and entrainment.[57] The United States is the only advanced nation whose practice of *infant circumcision* is both widespread and condoned by medical professionals.[58] Only about 16% of US hospitals are "baby-friendly" which refers to mostly *breastfeeding*-friendly practices (i.e., circumcision is not addressed), such as not separating mom and baby, not giving formula or sugar water.[59] The United States is the only advanced nation and one of the few in the whole world *without paid maternal leave after birth*, which with expectations of mothers to work and no child care support at work, undermines the evolutionarily "designed" mother-child relationship. Babies in the United States are often sent to child care centers where they are unlikely to receive the *affection*, carrying and *responsive care* needed.[60] U.S. cultural pressures to not "spoil the baby" or let baby "control you," eat away at mother's instincts to be responsive.[61] A mother must have great resolve to shut out this mistaken and harsh advice. A frequently distressed baby will develop stress-reactivity, a low threshold for getting upset, which they take forward into the rest of life.[62] Babies necessarily become distrustful of the world from the denial of their need for a mother's and allomothers' mothering. Through these (missing) practices, U.S. capitalism has prioritized money over family and relationships, breaking up extended family life, and undermining the feeling of connection children have toward the social world.

In all these ways and many others, the United States undermines the evolved nest, so we should not be surprised if its citizens display above average ill-being, aggression and antisocial behavior,[63] and an inability to feel connected to other peoples and other-than-humans. It should not be a surprise that a child undercared for would become an adult who reflects, believes in, and perpetuates the dominant Western worldview: that individuals are rational, self-contained, and autonomous, "locked within the privacy of a body," "standing against" and competing for the "rewards of success" with "an aggregate of other such individuals."[64] What else does the child know than feeling insecure in a dangerous, untrustworthy world?

The species-atypical human being, now widespread in most Western-industrialized societies, does not learn his place in the biocommunity because such learning was thwarted repeatedly in early life. The denial of the evolved nest teaches the child to deny himself, to distrust the world and to deny respect for others. It seems logical then to assume that a person raised outside the evolved nest feels "not of this world"—the world that rejected him so early. It is easy for him to believe that this is an alien place, and that there is someplace better. It is easy to adopt a mentality of "we are here only temporarily" and we will soon escape from the earth to something better. It takes a great deal of insight and healing to repair such early toxic damage.

Conclusion

How do we reenchant humanity with its heritages of living with the biocommunity and our evolved nest? Surely both are required for a sustainably wise life. The contributors to this volume remind us of what was and what we might do to return to sustainable wisdom. Jon Young points out the connectedness model that forms the basis of the original instructions of living together on the earth—with humans and other-than-humans.[65] The sense of connectedness is established at conception and, in traditional first nation societies, is maintained throughout life. It is a broad web of connections. Indigenous societies that existed for hundreds, if not thousands, of years around the world behave according to "what worked." They live intersubjective (relationally-attuned) lives with the particular landscape of plants, animals and other entities which teaches them how to live well in that place. They are adapted to living sustainably on the earth, following the laws of Mother Earth. These include the raising of a human being. They also follow the "original instructions" for raising human beings to live well on the earth.

How do we return to the heartminds that govern our sustainable ancestors? We build individual self confidence with the following types of practices. We provide the evolved nest to babies, children and adolescents, following ancient ways for becoming a full human being. We recognize, respect, and resonate with babies and young children as their brains and bodies construct themselves around such experiences. We help children learn to listen to and follow their heart of hearts instead of a set of rules or principles others have devised. We give people experience "in the wild" of their local landscape. We help children and adolescents apprentice into the important aspects of a good life: foraging, growing and cooking one's food; creative skills for making needed things (e.g., carpentry, sewing), alleviating the need for purchased items. We arrange children's lives around play and self-directed learning (Maria Montessori brought some of these ideas to schooling[66]). We provide extensive social support for people of all ages.

Thomas Berry, a noted spiritual father of the environmental movement, suggested that what is needed to transform human communities are not priests, intellectuals, or prophets—he actually pointed to education and religion as dominant failures of the 20th century.[67] Rationality has led humanity in the wrong direction and the cultural resources that depend on rationality have lost their integrity, leading instead to planetary destruction. Instead shamanic education is needed, by which he means returning to our human instincts, our wildness, our authentic spontaneity, from which a sustainable human culture will emerge. Our pre-rational resources will help us reinvent human culture. Our pre-rationality is fostered by the evolved nest, which includes close connection with the rest of the natural world—all our relations. Humans are embedded in human communities but also in unique landscapes of a biocommunity full of other-than-humans. Each place in the world has its own language, resonance, and ecological balance that influence how our biological needs are met and how our neurobiology functions. As enactive creatures, our psychobiosocial experiences influence all our capacities. We recognize our connections and responsibilities to the more-than-human. We apprentice one another into partnering with the needs of the local biocommunity. In this way we can return to being human and raising good human beings, using "original practices" that served us and our other-than-human relations for most of humanity's existence.

Discussion Questions

Describe the two worldviews. Which is most familiar to you? Which is most attractive to you?

What are the early experiences that help shape worldview (the evolved nest)?

Consider your own early experiences. How much did they follow the evolved nest?

Consider the experiences of young children in your community. How much do they follow the evolved nest?

Which community practices need to change to provide the evolved nest?

Go to this webpage for ideas of how to support evolved nest provision to young children: evolvednest.org

Notes

1. *Personal note*: It is difficult to hear and take in the viciousness of the recent centuries. I have a multicultural background, not only from experiences living around the world but with heritages from Germanic, Spanish, Jewish, Arab and Native peoples (Taino, we believe, since my father and his family were from Puerto Rico). Like all of us, my body has recollections of the trauma that our ancestors inflicted and received, passed to us through our parents (epigenetic inheritance). It is unclear to me why I am alive at this time when certain cultures of humanity are accelerating the destruction of manifest Life on the planet, except to help connect the dots on what contributes to such blind rapaciousness and suggest ways to prevent such madness in our children.

2. Sigmund Freud, *Civilization and its Discontents* (1929; reprint, London, UK: Penguin, 2002); Erich Fromm, *To Have or to Be?* (New York, NY: Harper and Row, 1976); Ashley Montagu, *Anthropology and Human Nature* (New York, NY: McGraw-Hill, 1963); Lewis Mumford, *Technics and Civilization* (1934; reprint, Chicago, IL: University of Chicago Press, 2010).

3. Freud, *Civilization and its Discontents.*

4. Marshall Sahlins, *The Western Illusion of Human Nature* (Chicago, IL: University of Chicago, 2008).

5. Robert Redfield, *The Primitive World and its Transformations* (Ithaca, NY: Cornell University Press, 1953); also his *Peasant Society and Culture: An Anthropological Approach to Civilization* (Chicago, IL: University of Chicago Press, 1956).

6. Frederick Turner, *Beyond Geography: The Western Spirit against the Wilderness* (1980; reprint, New Brunswick, NJ: Rutgers University Press, 1994).

7. Jeremy Lent, *The Patterning Instinct: A Cultural History of Humanity's Search for Meaning* (New York, NY: Prometheus Books, 2017).

8. Kirkpatrick Sale, *The Conquest of Paradise: Christopher Columbus and the Columbian Legacy* (New York, NY: Penguin Plume, 1990), 91.

9. William Woodruff, *Impact of Western Man: Study of Europe's Role in the World Economy* (Lanham, MD: Rowman & Littlefield, 1982), quoted in Sale, *The Conquest of Paradise*, 91.

10. Sale's words, quoted in Woodruff, *Impact of Western Man*, 88.

11. Daniel Quinn, *Ishmael: An Adventure of the Mind and Spirit* (New York: Bantam Book, 1992).

12. Term used in Turner, *Beyond Geography*.

13. Roy M. Robbins, *Our Landed Heritage: The Public Domain, 1776–1936* (Princeton, NJ: Princeton University Press, 1942); Turner, *Beyond Geography*.

14. Vine Deloria, *The World We Used to Live In: Remembering the Powers of the Medicine Men* (Golden, CO: Fulcrum, 2006); Tom Cooper, *A Time before Deception: Truth in Communication, Culture, and Ethics* (Santa Fe, NM: Clear Light, 1998).

15. Frank Bracho, "Happiness and Indigenous Wisdom in the History of the Americas," in *Unlearning the Language of Conquest*, ed. Four Arrows (Austin, TX: University of Texas Press, 2006), 42.

16. Turner, *Beyond Geography*, 129.

17. Ibid., 140.

18. N. B.: Like the Spanish, the English were also blind to the beauty and opportunity for spiritual and imaginative expansion that the New World offered. See Roxanne Dunbar-Ortiz, *An Indigenous People's History of the United States* (Boston, MA: Beacon Press, 2014).

19. Bob Blaisdell, *Great Speeches by Native Americans* (Mineola, NY: Dover Thrift Editions, 2000), 3.

20. Robert Berkhofer, *The White Man's Indian: Images of the American Indian from Columbus to the Present* (New York, NY: Alfred A. Knopf, 1978), quoted in Turner, *Beyond Geography*, 287.

21. Paul Shepard, *The Tender Carnivore and the Sacred Game* (New York, NY: Scribners, 1973).

22. R. D. Laing, *The Divided Self* (1959; reprint, London, UK: Penguin, 1990); David Shaw, *Traumatic Narcissism: Relational Systems of Subjugation* (New York, NY: Routledge, 2014); Donald Winnicott, *The Child and the Family* (London, UK: Tavistock, 1957).

23. Barry Hewlett and Michael Lamb, *Hunter-Gatherer Childhoods: Evolutionary, Developmental and Cultural Perspectives* (New Brunswick, NJ: Aldine Transaction, 2005), reviewed in Turner, *Beyond Geography*. Perhaps this is why thousands of European newcomers fled to live with the natives, or whom after capture and (forced) rescue, returned to live with natives at the earliest opportunity.

24. Gilda Morelli, Paula Ivey Henry, and Steffen Foerster, "Relationships and Resource Uncertainty: Cooperative Development of Efe Hunter-Gatherer Infants and Toddlers" in *Ancestral Landscapes in Human Evolution: Culture, Childrearing and Social Wellbeing*, ed. Darcia Narvaez et al. (New York, NY: Oxford University Press, 2014), 69–103.

25. John Mohawk, "From the First to the Last Bite: Learning from the Food Knowledge of Our Ancestors," in *Original Instructions: Indigenous Teachings for a Sustainable Future*, ed. Melissa K. Nelson (Rochester, VT: Little Bear & Co., 2008), 172.

26. Sarah Blaffer Hrdy, *Mothers and Others: The Evolutionary Origins of Mutual Understanding* (Cambridge, MA: Belknap Press, 2009).

27. Wenda Trevathan, *Human Birth: An Evolutionary Perspective*, 2nd ed. (New York, NY: Aldine de Gruyter, 2011).

28. William Greenough and James Black, "Induction of Brain Structure by Experience: Substrate for Cognitive Development," *The Minnesota Symposia on Child Psychology* 24 (1992):

155–200; Urie Bronfenbrenner, *The Ecology of Human Development* (Cambridge, MA: Harvard University Press, 1979); Hrdy, *Mother and Others.*

29. Gilbert Gottlieb, "On the Epigenetic Evolution of Species-Specific Perception: The Developmental Manifold Concept," *Cognitive Development* 17, no. 3–4 (2002): 1287–300; Mary Jane West-Eberhard, *Developmental Plasticity and Evolution* (New York, NY: Oxford University Press, 2003).

30. Melvin Konner, "Hunter-Gatherer Infancy and Childhood: The !Kung and Others," in *Hunter-Gatherer Childhoods: Evolutionary, Developmental and Cultural Perspectives,* ed. Barry S. Hewlett and Michael Lamb (New Brunswick, NJ: Transaction, 2005), 19–64.

31. Hrdy, *Mothers and Others;* Melvin Konner, *The Evolution of Childhood* (Cambridge, MA: Belknap Press, 2010).

32. Hewlett and Lamb, *Hunter-Gatherer Childhoods.*

33. Ibid., 15.

34. Darcia Narvaez, "Baselines for Virtue," in *Developing the Virtues: Integrating Perspectives,* ed. Julia Annas, Darcia Narvaez, and Nancy Snow (New York, NY: Oxford University Press, 2016).

35. Elysia Davis and Curt Sandman, "The Timing of Prenatal Exposure to Maternal Cortisol and Psychosocial Stress is Associated with Human Infant Cognitive Development," *Child Development* 81, no. 1 (2010): 131–48; Elysia Davis et al., "Prenatal Exposure to Maternal Depression and Cortisol Influences Infant Temperament," *Journal of the American Academy of Child and Adolescent Psychiatry* 46, no. 6 (2007): 737–46; Peter Gluckman and Mark Hanson, *The Fetal Matrix: Evolution, Development, and Disease* (New York, NY: Cambridge University Press, 2005).

36. Sarah J. Buckley, *Hormonal Physiology of Childbearing: Evidence and Implications for Women, Babies, and Maternity Care* (Washington, D.C.: National Partnership for Women and Families, 2015).

37. E.g., Stephen W. Porges, *The Polyvagal Theory: Neurophysiological Foundations of Emotions, Attachment, Communication, Self-Regulation* (New York, NY: W. W. Norton, 2011).

38. Armond S. Goldman, "The Immune System of Human Milk: Antimicrobial Anti-Inflammatory and Immunomodulating Properties," *Pediatric Infectious Disease Journal* 12, no. 8 (1993): 664–71.

39. Martin Reite and Tiffany Field, *The Psychobiology of Attachment and Separation* (Orlando, FL: Academic Press, 1985); Syndey Schanberg, "The Genetic Basis for Touch Effects," in *Touch in Early Development,* ed. Tiffany Field (Mahwah, NJ: Lawrence Erlbaum Associates, 1995); Myron. A. Hofer, "Early Social Relationships as Regulators of Infant Physiology and Behavior," *Child Development* 58, no. 3 (1981): 633–47.

40. Hrdy, *Mothers and Others.*

41. Gordon Burghardt, *The Genesis of Animal Play: Testing the Limits* (Cambridge, MA: MIT Press, 2005).

42. J. Burgdorf et al., "Uncovering the Molecular Basis of Positive Affect Using Rough-and-Tumble Play in Rats: A Role for Insulin-Like Growth Factor I," *Neuroscience* 168, no. 3 (2010): 769–77; Nakia S. Gordon et al., "Socially-Induced Brain 'Fertilization': Play Promotes Brain Derived Neurotrophic Factor Transcription in the Amygdala and Dorsolateral Frontal Cortex in Juvenile Rats," *Neuroscience Letters* 341, no. 1 (2003): 17–20;

Jaak Panksepp, "Can Play Diminish ADHD and Facilitate the Construction of the Social Brain?" *Journal of the Canadian Academy of Child and Adolescent Psychiatry* 16, no. 2 (2007).

43. Colwyn Trevarthen on play at birth: Colwyn Trevarthen, "Musicality and the Intrinsic Motive Pulse: Evidence from Human Psychobiology and Infant Communication," *Musicae Scientiae* 3, no. 1, suppl. (1999); and his "Communication Cooperation in Early Infancy: A Description of Primary Intersubjectivity," in *Before Speech: The Beginning of Human Communication*, ed. Margaret Bullowa (London: Cambridge University Press, 1979), 321–347.

44. Robin *Dunbar, Grooming, Gossip, and the Evolution of Language* (London, UK: Faber and Faber, 1996).

45. Allan N. Schore, "Effects of a Secure Attachment Relationship on Right Brain Development, Affect Regulation, and Infant Mental Health," *Infant Mental Health Journal* 22, no. 1–2, (2001): 7–66; and his *Affect Dysregulation & Disorders of the Self* (New York, NY: Norton, 2003).

46 Jill Bolke Taylor, *My Stroke of Insight* (New York, NY: Viking, 2008).

47. Process philosophical theories like those of Alfred North Whitehead, and dynamic psychological theories such as relational developmental systems theory typically are not integrated into mainstream research about the nature of the human being. For a critique that describes the latter theory, see David Witherington, Willis Overton, Robert Lickliter, Peter Marshall and Darcia Narvaez, "Metatheories and conceptual confusions in developmental science" *Human Development*, 61 (2018):181–198.

48. Paul Shepard, *Nature and Madness* (Athens, GA: University of Georgia Press, 1982); and his *Coming Home to the Pleistocene* (Washington, D.C.: Island Press, 1998), 7.

49. Jason Moore, *Capitalism in the Web of Life: Ecology and the Accumulation of Capital* (London, UK: Versa, 2015).

50. For reviews see Tim Ingold, "On the Social Relations of the Hunter-Gatherer Band," in *The Cambridge Encyclopedia of Hunters and Gatherers*, ed. Richard Lee and Richard Daly (New York: Cambridge University Press, 2005); Darcia Narvaez, "The 99%—Development and Socialization within an Evolutionary Context: Growing up to Become 'A Good and Useful Human Being,'" in *War, Peace, and Human Nature: The Convergence of Evolutionary and Cultural Views*, ed. Douglas Fry (New York, NY: Oxford University Press, 2013), 341–57.

51. Takashi Hayashi et al., "Laughter Up-Regulates the Genes Related to NK Cell Activity in Diabetes," *Biomedical Research* 28, no. 6, (2007): 281–85; Rosie Stacy, Katie Brittain, and Sandra Kerr, "Singing for Health: An Exploration of the Issues," *Health Education* 102, no. 4 (2002): 156–62; Elizabeth Valentine and Claire Evans, "The Effects of Solo Singing, Choral Singing and Swimming on Mood and Physiological Indices," *British Journal of Medical Psychology* 74, no. 1 (2001): 115–20; Norman Brown and Jaak Panksepp, "Low-Dose Naltrexone for Disease Prevention and Quality of Life," *Medical Hypotheses* 72, no. 3 (2009): 333–37; e.g., Daniel Leonard Everett, *Don't Sleep, There Are Snakes: Life and Language in the Amazonian Jungle*, 1st ed. (New York, NY: Pantheon Books, 2008).

52. Douglas P. Fry, *The Human Potential for Peace: An Anthropological Challenge to Assumptions about War and Violence* (New York, NY: Oxford University Press, 2006); Robert Knox Dentan, *The Semai: A Nonviolent People of Malaya* (New York, NY: Holt, Rinehart, and Winston, 1968).

53. E.g., Marc Bekoff, "Wild Justice and Fair Play: Cooperation, Forgiveness, and Morality in Animals," *Biology and Philosophy* 19, no. 4 (2004): 489–520; Sarah Brosnan, "Nonhuman Species' Reactions to Inequity and Their Implications for Fairness," *Social Justice Research* 19, no. 2 (2006): 153–85; Sarah F. Brosnan and Frans B. M. De Waal, "Monkeys Reject Unequal Pay," *Nature* 425, no. 6955 (2003): 297–99; Ernst Fehr, Helen Bernhard, and Bettina Rockenbach, "Egalitarianism in Young Children," *Nature* 454, no. 7208 (2008): 1079–83; Richard G. Wilkinson, "Socioeconomic Determinants of Health: Health Inequalities: Relative or Absolute Material Standards?" *British Medical Journal* 314, no. 7080 (1997): 591; For example, Inequality in social experience influences propensities toward drug use in monkeys; specifically, contemporaneous social experience such as ranking affects dopamine receptor binding and increases susceptibility to drug use in lower ranked monkey, see Drake Morgan et al., "Social Dominance in Monkeys: Dopamine D2 Receptors and Cocaine Self-Administration," *Nature Neuroscience* 5, no. 2 (2002): 169–74.

54. Harry F. Harlow, "The Nature of Love," *American Psychologist* 13, no. 12 (1958): 673–85.

55. Aida Gómez-Robles et al., "Relaxed Genetic Control of Cortical Organization in Human Brains Compared with Chimpanzees," *Proceedings of the National Academy of Sciences of the United States of America* 112, no. 48 (2015): 14799–804.

56. Charles Darwin, *The Descent of Man* (1871; reprint, Princeton, NJ: Princeton University Press, 1981). See Darcia Narvaez, "Evolution, Early Experience and Darwin's Moral Sense," in *Routledge Handbook of Evolution and Philosophy*, ed. Richard Joyce (London, UK: Routledge, 2017), 322–32.

57. Sarah J. Buckley, *Hormonal Physiology of Childbearing*; Thomas R. Verney, *Tomorrow's Baby: The Art and Science of Parenting from Conception through Infancy* (New York, NY: Simon & Schuster, 2002); Marsden Wagner, *Born in the USA How a Broken Maternity System Must Be Fixed to Put Mothers and Infants First* (Berkeley, CA: University of California Press, 2006); Gómez-Robles, "Relaxed Genetic Control."

58. See "Newborn Male Circumcision," American Academy of Pediatrics, accessed 30 July 2018, https://www.aap.org/en-us/about-the-aap/aap-press-room/pages/newborn-male-circumcision.aspx

59. See Baby-Friendly USA, "Upholding the Highest Standards of Infant Feeding Care," accessed 30 July 2018, https://www.babyfriendlyusa.org/.

60. Jay Belsky, "Developmental risks (still) associated with early child care," *Journal of Child Psychology and Psychiatry and Allied Disciplines. Oct*, (2001): 845–59. National Institute of Child Health and Human Development, *The NICHD study of early child care and youth development: Findings up to age 41/2 years* (Washington, DC: US Department of Health and Human Services, 2006).

61. Angel Braden and Darcia Narvaez, *Primal Parenting* (New York, NY: Oxford University Press, in press).

62. Sonia J. Lupien et al., "Effects of Stress throughout the Lifespan on the Brain, Behaviour and Cognition," *Nature Neuroscience* 10, no. 6 (2009): 434–45.

63. Organisation for Economic Cooperation and Development, *How's Life? 2013: Measuring Well-Being* (Paris: OECD, 2013), http://dx.doi.org/10.1787/9789264201392-en; Innocenti Research Centre, "An Overview of Child Well-Being in Rich Countries," *UNICEF* (2007), https://www.unicef-irc.org/publications/pdf/rc7_eng.pdf.

64. Ingold, "On the Social Relations of the Hunter-Gatherer Band," 407.
65. Jon Young, Ellen Haas, and Evan McGown, *Coyote's Guide to Connecting with Nature* (Santa Cruz: Owlink Media, 2010).
66. Maria Montessori, *The secret of childhood* (New York, NY: Ballantine Books, 1966).
67. Thomas Berry, *The Dream of the Earth* (San Francisco, CA: Sierra Club Books, 1988).

References

American Academy of Pediatrics. "Newborn Male Circumcision." Accessed 30 July 2018. https://www.aap.org/en-us/about-the-aap/aap-press-room/pages/newborn-male-circumcision.aspx.

Baby-Friendly USA. "Upholding the Highest Standards of Infant Feeding Care." Accessed 30 July 2018. https://www.babyfriendlyusa.org/.

Bekoff, Marc. "Wild Justice and Fair Play: Cooperation, Forgiveness, and Morality in Animals." *Biology and Philosophy* 19, no. 4 (2004): 489–520.

Belsky, Jay. "Developmental Risks (Still) Associated with Early Child Care." *Journal of Child Psychology and Psychiatry and Allied Disciplines.* Oct, (2001): 845–59.

Berkhofer, Robert. *The White Man's Indian: Images of the American Indian from Columbus to the Present.* New York, NY: Alfred A. Knopf, 1978.

Berry, Thomas. *The Dream of the Earth.* San Francisco, CA: Sierra Club Books, 1988.

Blaisdell, Bob. *Great Speeches by Native Americans.* Mineola, NY: Dover Thrift Editions, 2000.

Bracho, Frank. "Happiness and Indigenous Wisdom in the History of the Americas." In *Unlearning the Language of Conquest*, edited by Four Arrows, 29–44. Austin, TX: University of Texas Press, 2006.

Braden, Angela, and Darcia Narvaez. *Primal Parenting.* New York, NY: Oxford University Press (in press).

Bronfenbrenner, Urie. *The Ecology of Human Development.* Cambridge, MA: Harvard University Press, 1979.

Brosnan, Sarah F. "Nonhuman Species' Reactions to Inequity and Their Implications for Fairness." *Social Justice Research* 19, no. 2 (2006): 153–85.

Brosnan, Sarah F., and Frans B. M. De Waal. "Monkeys Reject Unequal Pay." *Nature* 425, no. 6955 (2003): 297–99.

Brown, Norman, and Jaak Panksepp. "Low-Dose Naltrexone for Disease Prevention and Quality of Life." *Medical Hypotheses* 72, no. 3 (2009): 333–37.

Buckley, Sarah J. *Hormonal Physiology of Childbearing: Evidence and Implications for Women, Babies, and Maternity Care.* Washington, D.C.: National Partnership for Women and Families, 2015.

Burgdorf, James Richard, R. A. Kroes, Margery C. Beinfeld, Jaak Panksepp, and Joseph R. Moskal. "Uncovering the Molecular Basis of Positive Affect Using Rough-and-Tumble Play in Rats: A Role for Insulin-Like Growth Factor I." *Neuroscience* 168, no. 3 (2010): 769–77.

Burghardt, Gordon. *The Genesis of Animal Play: Testing the Limits*. Cambridge, MA: MIT Press, 2005.

Cooper, Tom. *A Time before Deception: Truth in Communication, Culture, and Ethics*. Santa Fe, NM: Clear Light, 1998.

Darwin, Charles. *The Descent of Man*. 1871. Reprint, Princeton, NJ: Princeton University Press, 1981.

Davis, Elysia, and Curt Sandman. "The Timing of Prenatal Exposure to Maternal Cortisol and Psychosocial Stress is Associated with Human Infant Cognitive Development." *Child Development* 81, no. 1 (2010): 131–48.

Davis, Elysia, Laura M. Glynn, Christine Dunkel Schetter, Calvin Hobel, Aleksandra Chicz-Demet, and Curt Sandman. "Prenatal Exposure to Maternal Depression and Cortisol Influences Infant Temperament." *Journal of the American Academy of Child and Adolescent Psychiatry* 46, no. 6 (2007): 737–46.

Deloria, Vine. *The World We Used to Live In: Remembering the Powers of the Medicine Men*. Golden, CO: Fulcrum, 2006.

Dentan, Robert Knox. *The Semai: A Nonviolent People of Malaya*. New York, NY: Holt, Rinehart, and Winston, 1968.

Dunbar-Ortiz, Roxanne. *An Indigenous People's History of the United States*. Boston, MA: Beacon Press, 2014.

Dunbar, Robin. *Grooming, Gossip, and the Evolution of Language*. London, UK: Faber and Faber, 1996.

Everett, Daniel Leonard. *Don't Sleep, There Are Snakes: Life and Language in the Amazonian Jungle* (1st ed.). New York, NY: Pantheon Books, 2008.

Fehr, Ernst, Helen Bernhard, and Bettina Rockenbach. "Egalitarianism in Young Children." *Nature* 454, no. 7208 (2008): 1079–83.

Freud, Sigmund. *Civilization and its Discontents*. 1929. Reprint, London, UK: Penguin, 2002.

Fromm, Erich. *To Have or to Be?* (1st ed.). New York, NY: Harper & Row, 1976.

Fry, Douglas P. *The Human Potential for Peace: An Anthropological Challenge to Assumptions about War and Violence*. New York, NY: Oxford University Press, 2006.

Gluckman, Peter, and Mark Hanson. *The Fetal Matrix: Evolution, Development, and Disease*. New York, NY: Cambridge University Press, 2005.

Goldman, Armond S. "The Immune System of Human Milk: Antimicrobial Anti-Inflammatory and Immunomodulating Properties." *Pediatric Infectious Disease Journal* 12, no. 8 (1993): 664–71.

Gómez-Robles, Aida, William D. Hopkins, Steven J. Schapiro, and Chet C. Sherwood. "Relaxed Genetic Control of Cortical Organization in Human Brains Compared with Chimpanzees." *Proceedings of the National Academy of Sciences of the United States of America* 112, no. 48 (2015): 14799–804.

Gordon, Nakia S., Sharon Burke, Huda Akil, Stanley J. Watson, and Jaak Panksepp. "Socially-Induced Brain 'Fertilization': Play Promotes Brain Derived Neurotrophic Factor Transcription in the Amygdala and Dorsolateral Frontal Cortex in Juvenile Rats." *Neuroscience Letters* 341, no. 1 (2003): 17–20.

Gottlieb, Gilbert. "On the Epigenetic Evolution of Species-Specific Perception: The Developmental Manifold Concept." *Cognitive Development* 17, no. 3–4 (2002): 1287–300.

Greenough, William, and James Black. "Induction of Brain Structure by Experience: Substrate for Cognitive Development." *The Minnesota Symposia on Child Psychology* 24 (1992): 155–200.

Harlow, Harry F. "The Nature of Love." *American Psychologist* 13, no. 12 (1958): 673–85.

Hayashi, Takashi, Satoru Tsujii, Tadao Iburi, Tamiko Tamanaha, Keiko Yamagami, Rieko Ishibashi, Miyo Hori, Shigeko Sakamoto, Hitoshi Ishii, and Kazuo Murakami. "Laughter Up-Regulates the Genes Related to NK Cell Activity in Diabetes." *Biomedical Research* 28, no. 6 (2007): 281–85.

Hewlett, Barry, and Michael Lamb. *Hunter-Gatherer Childhoods: Evolutionary, Developmental and Cultural Perspectives.* New Brunswick, NJ: Aldine Transaction, 2005.

Hofer, Myron A. "Early Social Relationships as Regulators of Infant Physiology and Behavior." *Child Development* 58, no. 3 (1981): 633–47.

Hrdy, Sarah Blaffer. *Mothers and Others: The Evolutionary Origins of Mutual Understanding.* Cambridge, MA: Belknap Press, 2009.

Ingold, Tim. "On the Social Relations of the Hunter-Gatherer Band." In *The Cambridge Encyclopedia of Hunters and Gatherers*, edited by Richard Lee and Richard Daly, 399–410. New York, NY: Cambridge University Press, 2005.

Innocenti Research Centre. "An Overview of Child Well-Being in Rich Countries." *UNICEF* (2007). https://www.unicef-irc.org/publications/pdf/rc7_eng.pdf.

Konner, Melvin. "Hunter-Gatherer Infancy and Childhood: The !Kung and Others." In *Hunter-Gatherer Childhoods: Evolutionary, Developmental and Cultural Perspectives*, edited by Barry S. Hewlett and Michael Lamb, 19–64. New Brunswick, NJ: Transaction, 2005.

———. *The Evolution of Childhood.* Cambridge, MA: Belknap Press, 2010.

Laing, R. D. *The Divided Self.* 1959. Reprint, London, UK: Penguin, 1990.

Lent, Jeremy. *The Patterning Instinct: A Cultural History of Humanity's Search for Meaning.* New York, NY: Prometheus Books, 2017.

Lupien, Sonia J., Bruce S. McEwen, Megan R. Gunnar, and Christine Heim. "Effects of Stress throughout the Lifespan on the Brain, Behaviour and Cognition." *Nature Neuroscience* 10, no. 6 (2009): 434–45.

Mohawk, John. "From the First to the Last Bite: Learning from the Food Knowledge of Our Ancestors." In *Original Instructions: Indigenous Teachings for a Sustainable Future*, edited by Melissa K. Nelson, 170–79. Rochester, VT: Little Bear, 2008.

Montagu, Ashley. *Anthropology and Human Nature.* New York, NY: McGraw-Hill, 1963.

Montessori, Maria. *The Secret of Childhood* (New York, NY: Ballantine Books, 1966).

Moore, Jason. *Capitalism in the Web of Life: Ecology and the Accumulation of Capital.* London, UK: Versa, 2015.

Morelli, Gilda, Paula Ivey Henry, and Steffen Foerster. "Relationships and Resource Uncertainty: Cooperative Development of Efe Hunter-Gatherer Infants and Toddlers." In *Ancestral Landscapes in Human Evolution: Culture, Childrearing and Social Wellbeing*, edited by Darcia Narvaez, Kristin Valentino, Agustin Fuentes, James J. McKenna, and Peter Gray, 69–103. New York, NY: Oxford University Press, 2014.

Morgan, Drake, Kathleen A. Grant, H. Donald Gage, Robert H. Mach, Jay R. Kaplan, Osric Prioleau, Susan H. Nader, Nancy Buchheimer, Richard L. Ehrenkaufer, and Michael

A. Nader. "Social Dominance in Monkeys: Dopamine D2 Receptors and Cocaine Self-Administration." *Nature Neuroscience* 5, no. 2 (2002): 169–74.

Mumford, Lewis. *Technics and Civilization.* 1934. Reprint, Chicago, IL: University of Chicago Press, 2010.

Narvaez, Darcia. "Baselines for Virtue." In *Developing the Virtues: Integrating Perspectives,* edited by Julia Annas, Darcia Narvaez, and Nancy Snow, 14–33. New York, NY: Oxford University Press, 2016.

———. "Evolution, Early Experience, and Darwin's Moral Sense." In *Routledge Handbook of Evolution and Philosophy,* edited by Richard Joyce, 322–32. London, UK: Routledge, 2017.

———. "The 99%—Development and Socialization within an Evolutionary Context: Growing up to Become: 'A Good and Useful Human Being.'" In *War, Peace, and Human Nature: The Convergence of Evolutionary and Cultural Views,* edited by Douglas Fry, 341–57. New York, NY: Oxford University Press, 2013.

National Institute of Child Health and Human Development. *The NICHD study of early child care and youth development: Findings up to age 41/2 years* (Washington, DC: US Department of Health and Human Services, 2006).

Organisation for Economic Cooperation and Development. *How's Life? 2013: Measuring Well-Being.* Paris: OECD, 2013. http://dx.doi.org/10.1787/9789264201392-en.

Panksepp, Jaak. "Can Play Diminish ADHD and Facilitate the Construction of the Social Brain?" *Journal of the Canadian Academy of Child and Adolescent Psychiatry* 16, no. 2 (2007): 57–66.

Porges, Stephen W. *The Polyvagal Theory: Neurophysiological Foundations of Emotions, Attachment, Communication, Self-Regulation.* New York, NY: W. W. Norton, 2011.

Quinn, Daniel. *Ishmael: An Adventure of the Mind and Spirit.* New York, NY: Bantam Book, 1992.

Redfield, Robert. *Peasant Society and Culture: An Anthropological Approach to Civilization.* Chicago, IL: University of Chicago Press, 1956.

———. *The Primitive World and its Transformations.* Ithaca, NY: Cornell University Press, 1953.

Reite, Martin, and Tiffany Field. *The Psychobiology of Attachment and Separation.* Orlando, FL: Academic Press, 1985.

Sahlins, Marshall. *The Western Illusion of Human Nature.* Chicago, IL: University of Chicago, 2008.

Sale, Kirkpatrick. *The Conquest of Paradise: Christopher Columbus and the Columbian Legacy.* New York, NY: Plume, 1991.

Schanberg, Syndey. "The Genetic Basis for Touch Effects." In *Touch in Early Development,* edited by Tiffany Field, 67–80. Mahwah, NJ: Lawrence Erlbaum Associates, 1995.

Schore, Allan N. "Effects of a Secure Attachment Relationship on Right Brain Development, Affect Regulation, and Infant Mental Health." *Infant Mental Health Journal* 22, no. 1–2 (2001): 7–66.

———. *Affect Dysregulation & Disorders of the Self.* New York, NY: Norton, 2003.

Shaw, David. *Traumatic Narcissism: Relational Systems of Subjugation.* New York, NY: Routledge, 2014.

Shepard, Paul. *Coming Home to the Pleistocene*. Washington, D.C.: Island Press, 1998.

———. *Nature and Madness*. Athens, GA: University of Georgia Press, 1982.

———. *The Tender Carnivore and the Sacred Game*. New York, NY: Scribners, 1973.

Stacy, Rosie, Katie Brittain, and Sandra Kerr. "Singing for Health: An Exploration of the Issues." *Health Education* 102, no. 4 (2002): 156–62.

Taylor, Jill Bolke. *My Stroke of Insight*. New York, NY: Viking, 2008.

Trevarthen, Colwyn. "Communication Cooperation in Early Infancy: A Description of Primary Intersubjectivity." In *Before Speech: The Beginning of Human Communication*, edited by Margaret Bullowa, 321–47. London, UK: Cambridge University Press, 1979.

———. "Musicality and the Intrinsic Motive Pulse: Evidence from Human Psychobiology and Infant Communication." *Musicae Scientiae* 3, no. 1, suppl. (1999): 155–215.

Trevathan, Wenda. *Human Birth: An Evolutionary Perspective* (2nd ed.). New York, NY: Aldine de Gruyter, 2011.

Turner, Frederick. *Beyond Geography: The Western Spirit against the Wilderness*. 1980. Reprint, New Brunswick, NJ: Rutgers University Press, 1994.

Valentine, Elizabeth, and Claire Evans. "The Effects of Solo Singing, Choral Singing and Swimming on Mood and Physiological Indices." *British Journal of Medical Psychology* 74, no. 1 (2001): 115–20.

Verney, Thomas R. *Tomorrow's baby: The Art and Science of Parenting from Conception through Infancy*. New York, NY: Simon & Schuster, 2002.

Wagner, Marsden. *Born in the USA: How a Broken Maternity System Must Be Fixed to Put Mothers and Infants First*. Berkeley, CA: University of California Press, 2006.

West-Eberhard, Mary Jane. *Developmental Plasticity and Evolution*. New York, NY: Oxford University Press, 2003.

Wilkinson, Richard G. "Socioeconomic Determinants of Health. Health Inequalities: Relative or Absolute Material Standards?" *British Medical Journal* 314, no. 7080 (1997): 591–95.

Winnicott, Donald. *The Child and the Family*. London, UK: Tavistock, 1957.

Witherington, David, Willis Overton, Robert Lickliter, Peter Marshall and Darcia Narvaez, "Metatheories and Conceptual Confusions in Developmental Science." *Human Development*, 61 (2018):181–198.

Woodruff, William. *Impact of Western Man: Study of Europe's Role in the World Economy*. Lanham, MD: Rowman & Littlefield, 1982.

Young, Jon, Ellen Haas, and Evan McGown. *Coyote's Guide to Connecting with Nature*. Santa Cruz: Owlink Media, 2010.

Ways of Doing Science and Relating to Nature

CHAPTER SIX

Plants, Native Science and Indigenous Sustainability[1]

GREGORY A. CAJETE

Introduction

Plants present the life energy of the universe in their roots, stems, leaves, and flowers. In their tenacity for living in virtually every location on earth, plants exemplify the operation of the laws of nature, that of "life seeking life." At every turn, in every mode, and at every opportunity, plants seek to live their life and in their seeking to support all other life, including humans. Plants are an integral part of the earth's system of respiration. Plants are the primary living mechanisms for transforming and storing energy in forms that can be used by animals. Plants are essential partners in the evolution of life on earth since their appearance as tiny one-celled members of living communities over 2.5 billion years ago.

The intersection of plant and human nature was an integral consideration of Native science and Native societies, for they realized that a sustainable relationship with plants was the foundation of all human and animal life.[2] Why are plants so deeply embedded in the psychology of Native peoples? This deep relationship is rooted in the inherent focus of Native cultures on participation with Nature as the core thought and central dynamic of Native philosophy. The Native relationship to plants is also an expression of a universal human instinct to relate to a "green" nature. Plants present an internalized image of natural life and energy that helps to form our perception of the living earth. In reality, plants and humans have been biologically and energetically intertwined since the beginning of the human species. Our relationship to plants is a part of our body memory conditioned by the oldest survival instinct of humans. Our collective relationship to plants provides insights into the very core of human nature. In contemporary society this instinctual relationship and part of our nature is largely submerged beneath our intellect. Yet, this does not diminish the profound importance of plants and their central place in our psyche and physical being. It is no accident that human hemoglobin and plant chlorophyll share similar biochemical structures or that humans breathe oxygen produced by plant respiration and that plants depend on the carbon dioxide produced by humans and other animals.

In many Native myths, plants are acknowledged as the first life, or the grandparents of humans and animals and sources of life and wisdom, as in the case of the Native mythic symbol of the Tree of Life. Through such an acknowledgment of plants, Native myths mirror the reality of human biological evolution in the context of relationship to plants. For example, in the creation myth of the Inuit, the first man is born fully formed from a pea pod with the help of Raven, the Inuit trickster god. Plants profoundly influenced the development of the earliest human cultures. Indeed, the earliest evidence of human ritual activity as reflected in the contents of burial sites indicate that plants were used as food and medicine and in ceremony.

The origins of our instinctual body of knowledge and its preferences for certain qualities of nature form the foundation for our ability to survive as a species. Human preference for relationship to plants and for certain kinds of landscapes have evolved from our natural sense for those aspects of nature which have helped the human species survive. "Seeking life" is more than just a Native metaphor. It is an operational principle that is an extension of our instinctual predisposition for seeking life encoded in our genes. Green nature was the literal mother of our earliest human ancestors, providing food, clothing, shelter and a library of environmental "cues" for the kinds of landscape that could best support their "seeking of life."[3]

The philosophy of Native cultures focuses on a direct relationship with the earth as the source of knowledge and meaning for human life and community. In earth-centered traditions, each part of the earth is a manifestation of the spiritual center of the universe. Stones, trees, animals, or plants may be venerated as expressions of the sacred and representations of the greater family of life from which they come. Because plants are rooted in the earth and are intrinsically important to the life of humans, they are prime symbols for the life focus of Native science. Direct experience was the cornerstone of plant knowledge. Through experience, careful observation, and participation with plants, Native people came to possess a deep understanding of plant uses and relationship to humans, animals, and the landscape.

Plants were always viewed and utilized as a part of a greater context and were honored along with the sanctity of the natural place in which they grew. The cosmology of Native traditions often mentions plants, especially those of mind-altering qualities, as having special spirits that must be respected. Sacred plants and trees are mentioned in the guiding stories of Native traditions. Tobacco, corn, beans, squash, gourds, bitter root, peyote, mescal beans, and coca have all played significant roles in the lives, cultural development, and wellbeing of Native American peoples.

In the intimate relationships with their plants, Native people became sensitive to the fact that each has its own energy. "Coming to know," or understanding the essence of a plant derives from intuition, feeling, and relationship, and evolves over extensive experience and participation with green nature. This close relationship also leads to the realization that plants have their own destinies separate from humans, that is, Native people traditionally believed that plants have their own volition. Therefore, Native use of plants for food, medicine, clothing, shelter, art, and transportation, and as "spiritual partners," was predicated upon establishing both a personal and communal covenant with plants in general and with certain plants in particular.

Native uses of plants reflected their adaptation to varied environments. In the Southwest, plants grown were those adapted to arid environments, including corn, beans, squash, cotton, tobacco, and semi-wild plants such as devil's claw. In California, because of its diversity of environments, many varieties of plants were gathered, including acorns, pine nuts, and diverse seeds and grasses. In the Northwest, a rain forest ecosystem, the red cedar was by far the most important tree and a huge variety of plants species were used; cultures of the Northeast gathered plants including nuts, pot herbs, seeds, tubers, and roots, and used maple sap as a seasonal source of sugar. In the Plains region, wild plants such as prairie turnip and blue camas lily were gathered while agricultural was seasonally practiced, and many plants were gathered from the prairie turnip to the blue camas lily; crops included corn, beans, squash, tobacco and sunflowers. In the Southeast coastal areas, while fishing was predominant, an equally large variety of plants were gathered and supplemented with corn, beans, and varieties of squash and tobacco.

Native tribes throughout California practiced a kind of "environmental bonsai" through their centuries of hunting and gathering activities in that region. That is, Native practices of selective gathering in their prescribed ancestral territories actually formed the flora and fauna of their landscape. The harvesting of acorns, wild potatoes, pine nuts, buckeyes, bunch grass and other wild staples perpetuated these species and ensured their availability for people and animals.[4] Through the application of keen intellect, imagination, and a mythological sense of the diverse forms and functions of the plant world, Native cultures have evolved sophisticated ways of plant gathering, gardening, food preparation, and cooking that embody the essence of the participatory nature of Native science.

"Plants Are the Hair of Mother Earth"

> If I were to go for medicines I would first burn tobacco and tell the plants I was about to gather medicines. Then, all the plants would be ready for me to come. ... The tobacco is the medium of exchange that man has and with which he is able to procure the power of plants and animals; it is the vehicle of communication between men and all spiritual powers. This is oye gwa owe—the real tobacco of ongwe'owe, the real people.
>
> —Jesse Cornplanter[5]

Among some Indian herbalists, plants are referred to as "the hair of the earth Mother." There is a widespread traditional Native belief that the earth feels the pull every time a plant is taken from the soil. Therefore, humans must always make a proper offering and prayers. As one Navajo elder states, "You must ask permission of the plant or the medicine will not work. Plants are alive; you must give them a good talk."[6] Such offerings are made to ensure that the pulling of the "Earth Mother's hairs" does not hurt her too much and so that she understands that you comprehend your relationship to her and what she is giving you through a part of her body. In honoring and understanding this relationship, people also honor and understand their reciprocal relationship to all of life and nature.

The world of plants has spirit keepers. These plants might be the most frequently occurring types of trees in a particular environment, or they could be certain kinds of medicinal plants deemed important to a tribe. Native people have always understood that plants, like animals, have a quality of spirit that they share and which actually could be used to ensure the survival of a tribe. Therefore, ceremonies performed by tribes throughout North America incorporate symbolic representatives of plants and plant kingdoms. Ritual plants such as cornmeal, tobacco, and sweet grass are used as offerings to the spirit world and provided the material substance for both food and medicine. Moreover, plant symbols reflecting the sacred procreative power of earth abound in Native American philosophies.

The concept of the Tree of Life as a metaphor for a foundation of the cosmic order has been expressed in numerous ways in ceremony and philosophy throughout North America. Among the Pueblo people of the Southwest, the evergreen, which in most cases is the Western Fir or Blue Spruce, is a symbol of everlasting life and the connection of all life to the earth and the earth Mother. It is well understood that plants were the first kinds of living things and that both humans and animals depend on plant life for their existence. As the first living things, plants provide the most primal connection to the teeming life that is the most direct expression of Earth Mother's being.

Dependence on certain kinds of plants for the survival and maintenance of a people expresses itself in many ways. Among the Pueblo, corn, squash, pumpkin, and beans became the primary staple foods that gave rise to the social and community expressions. The relationship of the Pueblo farmer to corn is especially noteworthy. To Pueblo people, corn is a sacrament, a representation, and an embodiment of the essence of the earth Mother's life. Corn provided not only food but also a symbolic entity that cradled the entire psyche and spiritual orientation of the Pueblo peoples. Hopi farmers, in their fields at the foot of their mesas, developed practical technology for growing corn in inhospitable soil and, through long experience and comprehension of the growth of corn, evolved a variety of strains that grew well in different kinds of soils and environmental circumstances. Through understanding "the life and breath of corn," they established an elemental spiritual connection between themselves and this sacramental plant. That relationship extended not only into the technology of growing corn, but also into a variety of communal, artistic, and philosophical expressions.

Pueblo people evolved rituals and ceremonies that allowed them to express their partnership with those plants and animals. Among the Rio Grande Pueblos, Corn Dances performed during various times of the corn growing cycle reflect such communal expression. The community comes together—old and young, male and female—to celebrate and to provide occasion to "remember to remember" the connections that the community and individuals have with such a sacramental plant. This basic spiritual ecology, the understanding of Native people of their natural surroundings and the animals and plants upon which they depended, provided different occasions for expressing the relationship between themselves and the natural world. These expressions not only included the grand ceremonial representations of the plant and animal worlds, but were also reflected in engineering and other forms of practical agricultural and irrigation technologies. The great canals built in the Tucson valley by the Hohokam people long before Europeans came provide an early Native American example. Numerous other examples of this integration of technology with the practicality required to maintain a lifeway revolving around key plants—such as corn—occurred among the ancient Anasazi in southern Colorado. In each instance, the ceremony or communal work associated with corn occasioned people coming together to celebrate and to learn the nature of being in community and needing one another for survival.

Such instances of communal interaction with nature reflect a basic idea of natural community, of human beings who are active participants along with all other entities and energies within an environment. Traditional art forms reflect the attempt to understand human place within the natural community. The sense of natural community is expressed in the design motifs in art forms such as pottery. The Zia Pueblo sun, the horned serpent in black Santa Clara Pueblo pottery; representations of clouds, and of animals associated with water, such as the frog, water bird, dragonfly, tadpole, and butterfly, show the connection that Pueblo people felt with their environment. Such design motifs, inspired by natural entities and forces and the Pueblo peoples' experience of them, formed the basis of their artistic expressions and aesthetics. The commonly used motif of Corn Mother and her Corn Children, two perfect ears of corn each of a different color yet related through common ancestry, is another example of the metaphoric way that the intimate connections of plants, humans, animals, and all life were portrayed through Pueblo pottery.

For Pueblos, the making of pottery is a ceremonial act, an act of faith, an act of understanding the significance of relationship to the earth through bringing forth clay, as well as of reaffirming the basic connection that every human has to the earth. This sentiment, may be called "right relationship," also extends to the collection, preparation, and eating of food and foodstuffs from one's natural environment. What goes on or into the Pueblo pot is reflective of the realization that the food we eat must be appreciated—it is sacred and symbolic of that which gives us life and is a metaphor for our ultimate relationship with, and dependence upon, the natural world.[7]

Plants and the Foundations of Health and Wholeness

The food upon which Native people around the world depended for life was also their medicine. The two were so intimately intertwined that many foods, under proper supervision and application, were components of a medical system based on the natural properties of plants and animals. Food, combined with physical lifestyle and spiritual orientation, formed an interactive triad that was the cornerstone of health. In short, it is food that best symbolizes the ecology of Indigenous health.

Since all food that Native people ate came from the land or animals, it had a direct symbolic relationship to the way they viewed themselves *vis-a-vis* nature.

The place of food in ceremonies integrated their life experiences—intellectual, spiritual, and environmental. For instance, peoples of the Great Lakes domesticated and learned how to use the pond lily, wild rice, and other marsh-growing plants. These plants provided them not only with food but also a frame of reference for their existence and relationship to their place. Their knowledge and relationship with marshes was incorporated and reflected in their ability to make a living from their environment.

Native people learned how to use food plants occurring in their environment in the most productive, effective, and ecologically sound ways. In the Great Basin, the Paiutes evolved numerous uses for naturally-occurring pine nuts, a major source of protein. In the Northeast, the Algonquin peoples evolved a technology for using the sap of maple trees, which provided them not only with sugar, but also with a source of reflection and understanding of their connection to and dependence upon the trees. Through long experience, Native people came to know a plant's properties for healing certain illnesses. This knowledge, which ensures survival in a given environment, became an essential and basic foundation of Native education. As with all knowledge, how to use plants and animals for medicine evolved along a kind of continuum that was addressed in an appropriate way for each stage of life. Young people were given their first experience in how to use plants through simple observation and experience, through the use of story, and through actual demonstration.

This learning continuum evolved in a number of ways and incorporated into itself the full body of traditional philosophy and understanding of the human place within the greater natural world. Among all tribes, illness was associated with a kind of disharmony with some key element of the natural environment, so the healing rituals and ceremonies involved a re-establishment of harmony between the individual, family, or clan group and their immediate environment. They sought to realign with those energies and principles that formed expression of the group's lifeway. The teachers or mediators for the transfer of this knowledge of balance were primarily the healers or medicine people. Medicine people fulfilled a variety of roles. Herbalists, a principal group among the healers, were predominantly women whose work in many tribes was to gather wild foods. Through their experience with plant communities came an understanding of individual plants that could be used for medicine. Another group of healers had more knowledge, which might have included knowledge of the human body and muscular structure. They were adept at massage and repairing bones.

The next group were individuals, who had, as a result of their position, or clan initiation or societal membership, access to special knowledge of plants and animals. Through ceremonies, these individuals were empowered to address certain illnesses that they had learned to effectively treat.[8] In the words of Buhner, "To make the acquaintance of an herb, to understand the lowly weed, to hear its voice and that of the spirit teaching how to make it medicine and use it for healing, is the essence of earth relationship and earth healing—the essence of herbalism. ... It belongs to the realm where the human and the sacred meet in the plant."[9]

The most pronounced role of Native healing was that of the shaman. The shaman's knowledge of uses of plants for medicine embodied the most complete understanding of the nature of relationship between humans and the natural entities around them. These teachers provided a centering point for a teaching process that would result in establishing and maintaining balance between the community and the forces that acted upon it. Whether a singer in Navajo tradition, a tribal midwife, or a keeper of specific songs and dances among the Tlingit of the Northwest, the role of the shaman as First Doctor, First Visionary, First Dreamer, First Psychologist, First Teacher, and First Artist was focused on the spiritual ecology of the group in relation to its environment.

In the Native philosophy of healing, disease was always caused by improper relationship to the natural world, spirit world, community, and/or to one's own spirit and soul. Therefore, illness was always environmental. In one form or another, the environment provided the key context for illness, health, and the processes and expressions of healing that revolved around the re-establishment and maintenance of balance.

Breath was especially seen as being connected to the breath and the spirit of the earth itself. We breathe the same air that the plants breathe; we breathe the same air as animals; and we depend upon the same kinds of invisible elements as plants and animals. Therefore, we share a life of co-creation in an interrelated web of relationship that had to be understood, respected, and manipulated to maintain right relationships among important parts of the eco-system. Natural elements such as sun, fire, water, air, wind, snow, rain, mountains, lakes, rivers, trees, volcanoes, and a host of other entities played roles symbolically and physically in the expression and understanding of the ways of healing developed by Native people.

Among Native people, this understanding included all aspects of one's world and did not overlook the woven threads of the fabric of health. This is why Native people honor their heritage of knowledge and deeply appreciate that balance and harmony with the natural environment has to be maintained at all cost. Failure to do so would result in a cataclysm of dysfunction and disease due to environmental negligence.

Communal ceremonies were also tied directly into the guiding myth of a particular people and their self-understanding with the greater cosmos. Ceremonies were choreographed to help both individual and community come to terms one's relationship to other life. Among the Pueblos, the times of planting, growing, and harvesting of corn, are commemorated by the whole community. The Great Corn Dances of the Pueblos are timed with maturation of sacramental corn. The Great World Renewal ceremonies of the tribes of the Northwest and California, and the various Sun Dances of Plains peoples are renewal ceremonies in which the whole community participated. Traditional dance represents the personification of natural forces and entities—plants and animals—and natural phenomena in another highly evolved and sophisticated ceremonial expressions. The Winter Solstice ceremonies reflect constant vigilance in an annual cycle of celebration, ceremony, prayer, and teaching. The people represented their understanding of health and wholeness in metaphoric, symbolic, and ceremonial ways, a window through which to view the relationship they had established with the places where they lived.

Native Gardening

The instinctual connection to plants enters human experience in numerous ways through the windows, mirrors, and memories of our lives. For myself, growing up on the Santa Clara Pueblo Indian reservation in northern New Mexico, awareness of my personal connection to plants began at the hands of a master Pueblo gardener, my grandmother, Maria Cajete. Time spent in my grandmother's garden playing in the irrigation ditch eating green peas and melons while watching her lovingly care for each green plant instilled in me a special interest in and relationship with plants. Hearing her talk to each plant as if it were a person and hearing her scold away crows and my cat, "Tom," from her garden reflected her reverence for her garden. When I was older, she taught me how each plant needed certain things to encourage its growth. She told me how some plants like to grow in a family while others preferred to grow alone—just like people. And, just like people, she said, each plant had its own personality.

Pueblo gardening involved direct experiential learning about plants and their natural relationships to humans, to each other, and to their habitat. Children were guided both formally and informally in the technology of Native gardening. They were taught to distinguish among food plants that they were growing, useful herbs, and non-useful plants that could be pulled and used as mulch. Children's natural curiosity, intellect, and "biophilic" sensibility were developed simultaneously in their gardening activity. This early and direct involvement with the practical working of nature as reflect in gardening served to further develop Native children's capacity for participation with nature and the world of plants. In most traditions of

Pueblo gardening, cultivating domesticated food plants and promoting the growth of wild food plants are practiced simultaneously to maximize the availability of edible plants. Even insects that inhabited the garden might be used as added protein. The agricultural technologies of irrigation, soil preparation, fertilization, crop rotation, seed gathering, and storage were learned early by children. Certain plants had to be thinned or transplanted to ensure greater productivity. Pueblo gardens were constantly cared for to ensure that the relationship between the Pueblo farmer, the plants, community, and land was mutually beneficial to all. The Pueblo garden was a collaborative enterprise involving not only the individual farmer but the entire community and the land itself. All community members shared the tasks of cleaning the extensive network of irrigation ditches, and food preparation such as grinding corn and drying fruits.

Everyone in the community had a stake in their gardens. Every community member shared in the community garden's produce. Elders, women without spouses, and the disabled were provided for by the community. Sharing and giving of the produce of one's garden were Pueblo virtues and an expression of the interdependence of community, plants, and land. Yet, all of this has changed. In the words of San Juan Pueblo elder, Esther Martinez:

> The spirit of farming and the connection to nature in our Native communities is disappearing. With the modern ways of farming, one does not realize the value of the land and crops. Farming used to be done by families, who were involved with nature. It was good to get your hands all muddy and dirty, feeling and smelling the earth. We knew which areas were good for planting vegetable gardens and where to plant wheat, corn, and alfalfa. ... This absence of farming has affected our health in many ways. I don't remember our Indian people dying of cancer and a lot of the other problems that people have now. Working the fields kept us healthy and fit. It was a way of getting out to exercise. ... Nature was not only used to nourish our bodies but was utilized in all aspects of our lives. When my little brother and I were growing up, we gathered branches, old corncobs, and such. These were our toys. Nature provided for us all the things we needed for our survival; all we needed was a little imagination.[10]

"Native People Loved Their Gardens"

Through a combination of gathering and gardening, many Native tribes produced not only enough to live on but also a surplus for trade. Native people loved their gardens. Their gardens provided a place to socialize, hold gatherings and even a place to decorate. Gardens were an essential part of the ritual and ceremonial life of many communities as well. Plants such as corn, tobacco, amaranth, and quinoa were spiritually symbolic and a part of Native myths and cosmologies. The belief that humans were a part of the community of plants was widespread among many Native peoples. The following story of the Yuma Indians of the Sonoran desert relates how Brother Crow was responsible for bringing corn seeds to the people.

> Near the beginning of time Kakh, Brother Crow, brought to the people seeds of all sorts, especially corn, to nourish and replenish them. And the people cultivated their gardens so that when game was scarce they did not go hungry. Because Sister Corn gives of herself that they might live, the people return thanks in song and ceremonies at planting time and harvest.[11]

Stories like this of the coming of corn and its intimate relationship with people are numerous among the many Native groups who have cultivated it. Corn is after all a human-mediated plant, which means that it depends on humans for its cultivation and survival. This is an evolutionary relationship held in common with many domesticated plants and animals. The interdependence of humans and corn is a prime example of biological synergism that is often described metaphorically and imaginatively in Native myth.

The idea that human life is maintained through constant work, sharing, and relationship with food and other sources of life underlies the Native relationship to corn and other plants with which they have formed special reciprocal compacts. The Native garden provides an exemplification of this idea in an environmental and communal context of participation. When people eat the vegetables that grow in their gardens, the substance of the plants joins with the substance of the person in a way that is more than physical—more than survival of the body. It is a survival of the spirit also. The people's spirit also meets the spirits of the Corn Mother, or the Three Sisters, who give of their flesh to ensure the survival of the people.[12]

The Native garden involved a deep understanding of "practiced" relationship. Therefore, Native gardens were as much a mythic-spiritual-cultural-aesthetic expression of tribal participation and relationship with nature as was Native art, architecture, and ceremonialism. The technology of Native farming was only one dimension of such practiced relationship. This practiced relationship and responsibility to care for food plants such as corn is related in the following story of the Tuscarora Corn Spirit.

> In a time long past, the people of a village which had become known for its plentiful corn harvests became negligent of their compact with the Corn Spirit. The people had become used to the bountiful harvests and took corn for granted. They neglected to weed their gardens, store their seeds properly and give thanks to the corn spirit for their gardens. The people carelessly stored their corn and soon found that the mice had eaten their surplus. The men began to hunt but found that the game had also

vanished. Soon the people began to starve. Only one man named Dayohagwenda kept to the covenant of respect for the Corn Spirit. His garden continued to produce rich harvests. One day when Dayohagwenda was gathering herbs he came upon an old man dressed in rags and weeping. He asked the old man what had happened to which the old man replied, "Your people have forgotten me. I am the Corn Spirit and I will soon die." Dayohagwenda returned to the village and warned the people that the Corn Spirit might die if they did not start to once again honor him through carefully planting, weeding, harvesting, and storing their corn. They immediately began to do this and their corn once again began to grow in abundance as they gave thanks to the Corn Spirit for his blessing of food.[13]

Native cultures which practiced gardening evoked the spirits of plants and Nature to ensure or otherwise perpetuate the proper attitude and intentions for the success of their gardens. They asked that their gardens be protected from the ravages of Nature. In cultivating this attitude of reverence for their food plants, Native cultures expressed once again the central foundations of Native science— participation and relationship. Once planted, Native gardens were considered to have a life of their own and to be living expressions of the greater living community of the earth. They became living embodiments of the Earth Mother filled with spirit and in possession of a unique personality.

Native gardens also embodied skillful agricultural technology but involved Native spiritualism as well. In a metaphoric sense, the Native farmers "negotiated" with their gardens and with the spirits of Nature on their garden's behalf. They negotiated with the sun and rain for just the right amount of warmth and moisture to ensure their harvest. They negotiated with insects, birds and other animals on behalf of their gardens. Prayers and ritual were applied to request the good will of various other living energies which comprised the greater community in which the gardens were placed. When gardens did fail the blame could be placed not only on technical mistakes, but also on a loss of faith in them by the farmer, improper behavior during ceremonial observances, or errors in ritual practice.

When garden plants ripened in July and August ceremonies of thanksgiving were held. These rites of thanksgiving ranged from the Green Corn Dances of the Haida, Creeks and Choctaw to the Pueblo Corn Dances of the Southwest. Included in such ceremonial celebrations were countless other communal and personal acts of reverence which affirmed the connection to the greater dynamic of life at play in the Native garden. This awareness of connection did not exclude experimenting with the inherent possibilities of plants and the development of agricultural technologies. Native farmers were ingenious and practical scientists. Evidence of these traits are to be found in the domestication dozens of plants and the use of over two thousand plants for food by the various tribes of the Americas.[14] This made Native farming more productive and creative than farming in

Europe during the time of Columbus while at the same time practicing reverence and piety in relation to Green Nature.

For example, Machu Picchu may have actually functioned as an ancient experimental agricultural station. The Andeans probably did more plant experiments than any other people anywhere in the world.[15] But Native people lacked media exposure, and so their sophisticated agricultural techniques remained largely unknown until recently. These people experimented to improve crops with a vital purpose—in order to have enough to eat to survive, rather than in hopes of monetary profit or fame. Their careful scientific methods created very modern achievements that were remarkable on a small scale, and are as applicable today as in their times.

Western science has been reluctant to recognize Native peoples' agricultural methods, much less to acknowledge its debt to them. Permaculture techniques are being introduced to our children in schools today as newly developed ideas. Only now, when serious problems have resulted from large-scale monoculture, from insect invasion and pollution of food products grown in soil ruined and poisoned by pesticides, are agriculturists and healers looking back to the practical methods, knowledge, and philosophies used by Native peoples for many generations.

One might ask what Indigenous people today and of the recent past think about the contributions of their ancestors to the wealth and vitality of Western culture, which has spread throughout the world, while their own cultures have been undermined and their achievements ignored or treated with disdain.

> The roots of all living things are tied together. When a mighty tree is felled, a star falls from the sky. Before you cut down a mahogany, you should ask permission of the keeper of the forest, and you should ask permission of the keeper of the star.
> —Chan K'in of Naha,' *Last Lords of Palenque*.[16]

Creating New Models for Indigenous Development

There is a movement underway on the part of some Indigenous scholars to Indigenize foundational aspects of Indigenous development in ways that are more closely aligned with Indigenous world-views. In addition, this movement toward "indigenization" is tied to an evolving and increasingly more holistic and comprehensive approach to building Indigenous Nations. Recognizing the role of local Indigenous knowledge and creating infrastructures from the inside out based on inherent strengths with an eye toward "sustainability." are some of the key tenets of this movement toward indigenization. Indigenous people are learning, creating and evolving in their development of models for sustainability.

This creative process might be summarized as follows: gaining firsthand knowledge of community needs through "action research"—developing a comprehensive understanding of the history and "ecology" of a community economy—implementing strategies for regaining control of local economies—creating models based on lessons learned and application of research of practices which work—cultivating networks for mutual support. This new movement and new thinking regarding Indigenous development is in direct contrast with the standard approaches of the past, which mimic the Western mono-dimensional model of development. Indeed, the underlying assumptions, aims, and effect of the Western model must be questioned in terms of their ultimate sustainability. It is through the application of the lens of sustainability that Indigenous people have come to realize the wisdom and consequences of applying the Western model of development to their circumstances. It is also with the application of the conceptual framework of sustainability that the greatest opportunity of the application and even evolution of Indigenous Science as a living and evolving base of knowledge upon which Indigenous communities might rely.

In using "*community sustainability*" as a guiding paradigm for building Indigenous Nations, the underlying assumptions and mono-dimensionality of the standard Western development model become apparent. Upon close examination, the limitations of the Western development paradigm in helping Indigenous communities realize their goals of empowerment, renewal and revitalization are also apparent.

The Philosophical Base of Native Science

Development of knowledge through Native science is guided by spirituality, ethical relationship, mutualism, reciprocity, respect, restraint, a focus on harmony, and acknowledgement of interdependence. This knowledge is integrated with regard to a particular "place" toward the goal of sustainability. Native science knowledge is derived using many of the same methods as modern Western science, including: classifying, inferring, questioning, observation, interpreting, predicting, monitoring, problem solving and adapting. The difference is that Native science perceives from a "high context" view including all relational connections in its consideration. In contrast, Western science perceives from a "low-context" view, reducing awareness of complex interconnections to a minimum.

Native science may be defined as a "multi-contextual" system of thought, action and orientation applied by an Indigenous people through which they interpret how Nature works in "their place." *Native knowledge* may be defined as a "high-context" body of knowledge built up over generations by culturally distinct people living in close contact with a "place," its plants, animals, waters, mountains, deserts, plains, etc. Epistemological characteristics of Native science include oral transmission, observation over generations, cyclic time orientation, quantification at a macro level, a specific cultural/literary style, and symbolism. Knowledge is contexted to a specific tribal culture and place and is conserved through time and generations.

Thus, Native science and culturally-responsive education support the inclusion of Indigenous knowledge *on an equal par* with modern Western science. This is a relatively new and radical idea for Western science which has been met with much debate. Proponents of inclusion of Indigenous science argue that all cultures have developed a form of science which is important to the overall diversity of human knowledge related to the biosphere. However, for some only Western science is "true science" and all other forms of knowledge must be subordinate. Despite such attitudes, teaching for sustainability can provide a context for the inclusion of Indigenous science in all aspects of science education.

Sustainability, or the ability of current and future generations to meet their basic needs within the context of community, is by its nature an interdisciplinary inquiry that is inclusive of sciences, technology, business, politics, philosophy and the arts. This inquiry takes place with focus upon a specific place, populations, and time period. The goals of such an inquiry are to engage students in the *production of knowledge*, to learn various research methods, to develop a critical voice in writing, and most importantly, understand the importance of sustainability. According to David Orr, there are four challenges to pursuing education for sustainability. These are: (1) creating better, more integrated science and accounting tools to measure biophysical wealth; (2) getting people involved; (3) transforming societal value systems through "empathic education"; and (4) improving knowledge transfer around sustainability.[17] Tied to these challenges are associated issues revolving around human health, social justice, equity, economic development, ethical valuing and governance. The context of relationships in which this occurs must bring about the balanced and ethical interaction of three interacting circles: individuals, community and the environment. And in these understandings and relationships, the aim must be to maintain cultural diversity, protect human health, create and maintain sustainable economic relationships, reconcile social issues non-violently, and, most essentially, protect the environmental life support system.

In conclusion, *Native Science Education* can be strategically applied to educate for the re-creation of cultural economies around an Indigenous paradigm of sustainability. This begins by learning the history of a particular Indigenous way of sustainability and exploring ways to translate its principles into the present.

There must be research into the practical ways to apply these Indigenous principles and knowledge base. Added to all this, Indigenous peoples must revitalize, re-learn, or otherwise maintain their traditional environmental knowledge. This can be accomplished through application of the Indigenous communal strengths of resourcefulness, industriousness, collaboration, and cooperation. In addition, we must once again apply our collective and historical ability to integrate differences in our political organizations, forge alliances and confederations, and re-introduce our propensity for trade and exchange. We have ancient systems of extended family, clan, and tribal relationship that we can mobilize in positive ways to implement sustainable changes in our economies. In addition, we have developed modern political, social, professional trade organizations, federations, associations and societies which we can enlist in the addressing the challenges which we now collectively face. These are the new areas of Indigenous education which must be explored and operationalized in the context of Indigenous education toward the development and re-vitalization of Native communities as they face the challenges of surviving the ecological, social and political challenges of a 21st century world.

Discussion Questions

How is Native science different from the dominant form of science today?

How do Native peoples treat food and how is that different from the way food is treated in modern industrialized societies?

How are education practices in Native societies different from those in modern industrialized societies?

Describe the relationship with gardens and gardening that Native peoples have.

Notes

1. Portions of this chapter have been adapted from a previously published work: Gregory A. Cajete, *Native Science: Natural Laws of Interdependence* (Santa Fe, NM: Clear Light, 2000), 106–47.
2. The terms *Native, Indigenous, Tribal,* and *Tribe* are capitalized to add emphasis and to convey an active and evolving identity. (The terms *Native* or *Indigenous* are used as the larger inclusive group term while *Tribal* refers to specific contexts, these terms are capitalized as an honorific designation. American Indian is used when referring specifically to a tribe that resides in the United States.
3. Charles A. Lewis, *Green Nature, Human Nature* (Urbana, IL: University of Illinois Press, 1996), 1–13.

4. Dennis Martinez, "Conference Remarks," (presented at the Law and Theology Conference of the American Indian Science and Engineering Association, Boulder, CO, 12–13 June 1992).

5. This quotation of Jesse Cornplanter found in James William Herrick, "Iroquois Medical Botany." (Ph.D. diss., State University of New York at Albany, 1977), 136–37.

6. Donald Hughes, *American Indian Ecology* (El Paso, TX: Texas Western Press, 1983), 64.

7. Gregory A. Cajete, *Look to the Mountain: An Ecology of Indigenous Education* (Skyland, NC: Kivaki Press, 1994), 100–01.

8. Ibid., 105–06.

9. Stephan Harrod Buhner, *Sacred Plant Medicine: Explorations in the Practice of Indigenous Herbalism* (Boulder, CO: Roberts Rhinehart, 1996), 101.

10. Quoted in Gregory A. Cajete, ed., *A People's Ecology: Explorations in Sustainable Living* (Santa Fe: Clear Light, 1999). 129.

11. Carol Buchanan, *Brother Crow, Sister Corn: Traditional American Indian Gardening* (Berkeley, CA: Ten Speed Press, 1997), 5.

12. Buchanan, *Brother Crow, Sister Corn*, 7

13. Adapted from "The Corn Spirit" in Tom Lowenstein and Piers Vitebsky, *Mother Earth, Father Sky* (Amsterdam: Time-Life Books, 1997).

14. Edwin Francis Walker, *World Crops Derived from Indians* (Los Angeles, CA: The Southwest Museum, 1943).

15. Jack Weatherford, *Indian Givers: How Indians of the Americas Transformed the World* (New York, NY: Ballantine Books, 1988), 62.

16. Ibid., 62.

17. David Orr, *Earth in Mind: On Education, Environment, and the Human Prospect* (Washington, D.C.: Island Press, 1994).

References

Buchanan, Carol. *Brother Crow, Sister Corn: Traditional American Indian Gardening.* Berkeley, CA: Ten Speed Press, 1997.

Buhner, Stephan Harrod. *Sacred Plant Medicine: Explorations in the Practice of Indigenous Herbalism.* Boulder, CO: Roberts Rhinehart, 1996.

Cajete, Gregory A., ed. *Farming the Memories.* Santa Fe, NM: Pueblo of Pojoaque Poeh Museum, 1995.

———. *Look to the Mountain: An Ecology of Indigenous Education.* Skyland, NC: Kivaki Press, 1994.

Herrick, James William. "Iroquois Medical Botany." Ph.D. diss. State University of New York at Albany, 1977.

Hughes, Donald. *American Indian Ecology.* El Paso, TX: Texas Western Press, 1983.

Lewis, Charles A. *Green Nature, Human Nature.* Urbana, IL: University of Illinois Press, 1996.

Lowenstein, Tom and Piers Vitebsky. *Mother Earth, Father Sky.* Amsterdam: Time-Life Books, 1997.

Martinez, Dennis. "Conference Remarks." Presented at the Law and Theology Conference of the American Indian Science and Engineering Association, Boulder, CO, 12–13 June 1992.

Orr, David. *Earth in Mind: On Education, Environment, and the Human Prospect*. Washington, D.C.: Island Press, 1994.

Walker, Edwin Francis. *World Crops Derived from Indians*. Los Angeles, CA: The Southwest Museum, 1943.

Weatherford, Jack. *Indian Givers: How Indians of the Americas Transformed the World*. New York, NY: Ballantine Books, 1988.

CHAPTER SEVEN

Mother Earth vs. Mother Lode

Native Environmental Ethos, Sustainability, and Human Survival[1]

BRUCE E. JOHANSEN

The development of an adequate environmental ethos in modern times reflects Native wisdom and values, as well as recognition that traditional capitalism is a suicide pact; as Edward Abbey said decades ago, "the ideology of the cancer cell" is random, all-consuming, and eventually lethal to its host. I will trace the development of this ethos and its adaptation in non-Native society as necessary for human survival. My primary exhibit of this adaptation will be increasing awareness of global warming and its intensifying perils for coming generations. Specifically, I will focus on the geophysical principle of thermal inertia, which describes how the system delivers evidence of temperature rise 50 years (in the air) and about 150 years (in the oceans) after the carbon-dioxide emissions that cause them. The geophysical system thus requires that our industrial and diplomatic systems respond according to the needs of the seventh generation, in accordance with Native traditional ecological ethos.

The idea that Native peoples have something valuable to teach majority society has become quite popular. When I proposed to trace an eighteenth-century tutorial by the Haudenosaunee (Iroquois) for Benjamin Franklin forty years ago, most of my Ph.D. dissertation supervisors thought I was, if not crazy, at least out of paradigm. Four decades after I started my dissertation journey, during mid-December 2015, I woke up in South India, having been invited to give a plenary address: "What Has Been Will Be: Native American Contributions to Democracy,

Feminism, Gender Fluidity, and Environmentalism" at a Global Seminar on "Celebrating the Ancient/Contemporary Wisdom of Fourth World" hosted by the Department of English, Acharya Nagarjuna University, Guntur, India.

I was arguing that the "history of westward movement," as it was parsed at that time, was much more nuanced than that. Discovery goes both ways in any encounter. So while Europe did not "discover" America, it *was* quite a discovery for Europe. For roughly three centuries before the American Revolution, the ideas that made it possible were being discovered, nurtured and embellished in the growing English and French colonies of North America, as images of America became a staple of European literature and philosophy. America provided a counterpoint for European convention and assumption. America became, for Europe and Europeans in America at once a dream and a reality, a fact and a fantasy, the real and the ideal.

A detailed case has now been made that Native consensual democracy helped shape the thoughts of some of the United States' founders. Decades after that, the founding mothers of American Feminism learned from Native matrilineal cultures. Today (witness the recent recognition of gay marriage by the U.S. Supreme Court), Indigenous American acceptance of gender fluidity has become accepted as well. Environmental points of view (that all things are connected, for example) also have entered mainstream thought, but only after the perils of industrial pollution and alteration of the atmosphere (climate change) have become manifest.

Environmental Ethos: Mother Earth or Mother Lode?

Native American philosophy often combines spiritual and environmental themes in ways that appeal to many non-Indian environmental activists today. A lively scholarly debate has flared regarding how Native Americans generally conceived of the Earth. Some ethno-historians maintain that Native Americans possessed little or no environmental philosophy, and that any attempt to assemble evidence to sustain a Native American ecological paradigm is doomed to failure because the entire argument is an exercise in wishful thinking by environmental activists seeking sentimental support for their own views.

One wishes to be intellectually charitable, and not to unfairly alienate William A. Starna and others who share beliefs about the recent, non-Indian genesis of the "mother earth" image. However, their assertions that Native Americans had no concept of mother earth before immigrant peoples fantasized it in their name during the twentieth-century environmental movement misses an astonishing amount of the historical record, not to mention Native American oral histories. References to "mother earth" in Native American cosmology are not scarce. They are abundant, according to George Cornell: "Native peoples almost universally view the earth as a feminine figure. The Mother provides for the sustenance and wellbeing of her children: it is from her that all subsistence is drawn. The relationship of native peoples to the earth, their Mother, is a sacred bond with the creation."[2]

Many Native cosmologies conceive of the sky (including the sun) as a masculine counterpart to mother earth, as a loving couple who are sometimes prone to many of the failings of human relationships between men and women. While *The Bible* in *the book of Genesis* commands human beings to subdue the Earth (a phrase recently interpreted by Catholic theologians to mean act as steward), humankind in many Native cosmologies places human beings in a web of interdependent relationships with all facets of the Creation. In this web, all things are animate, even objects, such as the pebbles under one's feet, which European languages characterize as lifeless. In the web of Native American experience, the landscape of life envelops *all* of reality.

European-Americans have been hearing Native Americans characterize the earth as mother since shortly after the Mayflower landed. Massasoit, who invited the Pilgrims to the first Pilgrim Thanksgiving dinner, faced European ideas of land tenure with a few questions of his own: "What is this you call property? It cannot be the earth, for the land is our mother, nourishing all her children, beasts, birds, fish, and all men. The woods, the streams, everything on it belongs to everybody and is for the use of all. How can one man say it belongs only to him?"[3]

Those who dismiss a Native ecological ethic as the invention of modern-day hippies and pan-Indianists are missing something much deeper than mere mentions of "mother earth" in nineteenth-century primary sources. They are missing the fundamental nature of many Native American traditions, the terms in which Native thought conceptualizes the land and the life it nurtures. To Roger Dunsmore, Western thought creates hierarchies and categories that do not exist in Native American thoughtways.[4] The very cognitive map for conceptualizing life is different, as illustrated in recent time by the example of "... A Wasco Indian logger (a faller), who quit logging and sold his chainsaw because he couldn't stand hearing the trees scream as he cut into them."[5] This is a worldview in which "[T]he whole world is perceived and valued. Even the flies."[6] A sense of a web of life connecting all things framed the cognitive map of Chief Joseph when he said, "The Earth and I are of one mind."[7]

Sometimes a belief in earth as mother is reflected in Native American languages. In the Algonkian Ojibwe language, for example, "The words for Earth and the vagina, respectively *aki* and *akitun*, share the same root."[8] Each Native people in the Americas has its own origin story, but many share common elements. The characterization of the earth in the feminine, using kin terminology, is one of these. Native American perspectives on the environment often were virtually opposite to those of many early settlers, who sought to "tame" the "wilderness." Many Native Americans saw themselves as enmeshed in a web of mutually complementary life. As Black Elk said, "With all beings and all things, we shall be as relatives."[9]

References to Indigenous affection for nature permeated the thoughts of Luther Standing Bear, who wrote:

> The Lakota was a true naturalist—a lover of Nature. He loved the earth and all things of the earth, the attachment growing with age. The old people came literally to love the soil and they sat or reclined on the ground with a feeling of being close to a mothering power. ... In talking to children, the old Lakota would place a hand on the ground and explain: "We sit in the lap of our mother. From her, we, and all other living things, come. We shall soon pass, but the place where we now rest will last forever ... Our altars were built on the ground and were altars of thankfulness and gratefulness. They were made of sacred earth and placed upon the holiest of all places—the lap of Mother Earth.[10]

Standing Bear defined his people's relationship to everything else on Earth, writing that in the native view everything is animate—"possessed of personality." He compared the world to a library, with "the stones, leaves, grass, brooks ... birds, and animals as its books."[11] Many times, wrote Standing Bear, "The Indian is embarrassed and baffled by the white man's alienation from nature, as reflected in allusions to nature in such terms as "crude, primitive, wild, rude, untamed, and savage."[12] To Standing Bear, many whites imagined Native Americans as savages to "salve ... [their] sore and troubled conscience[s] now hardened through the habitual practice of injustice."[13]

Standing Bear, who watched large-scale Anglo-American immigration change the face of the Great Plains, contrasted European-American and Native American conceptions of the natural world of North America:

> We did not think of the great open plains, the beautiful rolling hills, and winding streams with tangled brush, as "wild." Only to the white man was nature "a wilderness" and only to him was the land "infested" with "wild" animals and "savage" people. To us it was tame. Earth was bountiful, and we are surrounded with the blessings of the Great Mystery. Not until the hairy man from the east came and with brutal frenzy heaped injustices upon us and the families we loved was it "wild" for us. When the very animals of the forest began fleeing from his approach, then it was for us that the "Wild West" began.[14]

Standing Bear was a severe critic of the whites' attitudes toward nature. He said he knew of no species of plant, bird, or animal that had been exterminated in America until the coming of the white man. For some years after the buffalo disappeared, there still remained huge herds of antelope, but the hunter's work was no sooner done in the destruction of the buffalo than his attention was attracted toward the deer. "The white man considered natural animal life just as he did natural [Native American] life upon this continent, as 'pests,'" wrote Standing Bear. "Plants which the Indian found beneficial were also 'pests.' There is no word in the Lakota vocabulary with the English meaning of this word."[15]

Like Black Elk, Tecumseh, Black Hawk, and others, Standing Bear invoked the image of mother earth in his writing. Luther Standing Bear's use of the earth-mother image is particularly striking when placed next to similar language used by Black Elk which has been passed to us through accounts by Neihardt, Epes-Brown, and others. Unlike accounts attributed to Black Elk, Tecumseh and Sea'thl, however, the use of the image by Standing Bear raises no questions of interpretation, because he wrote in English acquired at the Carlisle Indian School.

> There is a great difference in the attitude taken by the Indian and the Caucasian toward nature, and this difference made of one a conservationist and the other a non-conservationist of life. The Indian, as well as other creatures that were given birth, were sustained by the common mother—earth. He was therefore kin to all living things and he gave to all creatures equal rights with himself ... The ... Caucasian. ... Bestowing upon himself the position and title of a superior creature, others in the scheme were, in the natural order of things, of inferior position and title; and this attitude dominated his actions toward all things. The worth and right to live were his, thus he heartlessly destroyed. Forests were mowed down, the buffalo exterminated, the beaver driven to extinction and his wonderfully constructed dams dynamited ... the white man has come to be the symbol of extinction for all things natural to this continent.[16]

Standing Bear was also a critic of European-American society generally, in words appreciating nature and his forefathers and foremothers that recall those of Sea'thl (Seattle), three-quarters of a century earlier. Standing Bear also evoked the same sacred tree of life that was familiar to Black Elk:

> The white man does not understand the Indian for the reason that he does not understand America. He is too far removed from its formative processes. The roots of the tree of his life have not yet grasped the rock and soil. The white man is still troubled with primitive fears; he still has in his consciousness the perils of this frontier continent. ... He shudders still with the memory of the loss of his forefathers upon its scorching deserts and forbidding mountain-tops. The man from Europe is still a foreigner and an alien. And he still hates the man who questioned his path across the continent.[17]

An Environmental Ethos

In addition to being rooted in their homelands, Native Americans maintain historical, spiritual bonds to the land that foster attention to environmental threats. Long-time fishing-rights activist Billy Frank, Jr. said that this connection places protection of environment and its role in sustaining human and all other life "at the top of our priority list."[18] "When we say the Okanagan [Native] word for 'ourselves,'" said Jeanette Armstrong, "We are actually saying the ones [that] are 'dream' and 'land' together."[19] Chief Willie Charlie has said: "Mother Earth is crying, and we need to pay attention to what she is saying."[20]

Activist and author Kurt Russo, who works with the Native American Land Conservancy, commented:

> Our courage to acknowledge this crisis, our conviction to stand up for unborn generations, our connection to nature and, through nature, to each other, and our resilience—as a family, as a species, as peoples—will determine whether we hear her cry for the agony of extinction or stand idly by and bear witness to a great dying.21

He also noted:

> Indigenous communities are, in general, in a unique position given their history and knowledge to understand and respond to the crisis. [They] are, in general, more informed and engaged than the majority of Americans or their political and corporate leaders. ... As place-based communities of inter-related families with historical consciousness, Indigenous peoples are also more resilient, and thus able to face, rather than deflect or deny, the true magnitude of the crisis [that] will have long-lasting and potentially catastrophic consequences for every life-form and human community.[22]

Daniel Wildcat, Yuchi member of the Muscogee Nation of Oklahoma and a professor at Haskell Indian Nations University, Lawrence, Kansas, speaking at the University of Colorado-Boulder's Center of the American West on September 29, 2011, said that, until recently, Native people were members of tribes, not nation-states, with a relationship to nature that defined species on which people depended for survival in familial terms, as relatives, not as exploitable resources. The bison on the Great Plains, salmon among the Coast Salish, and corn across much of today's North America (Turtle Island), was "the central relative we acknowledged."[23]

Indigenous environmental activism (including visceral opposition to development of extractive industry) stems from long historical experience with resource colonization, which works in synthesis with a spiritual ethos that invests animus in everything natural. While European religions often restrict their blessings to humanity, many Native Americans interpret "all my relations" to mean *all* of nature—animate and not. This respect for nature is fundamental and enduring, and at the root of traditional Native American responses to economic development. Definitions of "balance" are couched in this context, counterpoising protection of "mother earth" with an immigrants' ethos that seeks a "mother lode," rooted in *Genesis* 1:28 ("Be fruitful, multiply, fill the Earth and subdue it"). Naomi Klein, in *This Changes Everything: Capitalism and the Climate* writes of industrialists who "view … nature as a bottomless vending machine."[24]

Global Warming and Natural Limits

Understanding how natural systems operate is at the center of mother-Earth thinking. What follows (on thermal inertia and climate science) may seem "off topic" at first glance, because it crosses disciplinary boundaries. It is really at the heart of the matter. Such understanding is vital to a realization of how vital a paradigm change in our belief system contributes to a sustainable future. In this sense, we are not "off topic." Do not let categorical thinking obscure a realization that, in this case, everything is connected.

Today, the most pervasive and urgent illustration of mother-lode thinking's unsustainable legacy is global warming, as levels of greenhouse gases rise to dangerous levels. The fossil fuel age dawned just as the United States became the Earth's most powerful economy, built across an expanding territory with surging immigration (mainly, but not entirely, from Europe). Exploitation of coal, followed by oil and natural gas between the mid-nineteenth and early twentieth centuries introduced machine labor representing the equivalent of a billion horses (or 3 billion human slaves). Not coincidentally, perhaps, human slavery became economically as well as politically obsolete. As an illustration of just how much human labor was transferred to fossil-fueled machines between 1800 and 1970, consider that the number of human hours of labor going into an acre of wheat declined from 56 to 2.9. For an acre of cotton, the same figure declined from 185 to 24. The raising of food has become as mechanized as the manufacture of anything else: seven calories of energy (mainly fossil fuels) by 2014 was required to produce one calorie of food.[25] This revolution in energy generation increased production of heat-retaining greenhouse gases in Earth's atmosphere.

As part of Earth's natural cycle, the greenhouse effect is very necessary to life on Earth. Without it, the planet's average temperature would be around zero degrees Fahrenheit. It is the added warming provoked by human combustion of fossil fuels that causes a problem. Like chocolate, a little is a good thing; too much is toxic to the system. Fossil fuels provide us comfort and convenience, and altering their use in a fundamental way presents the challenge of the century. Unless we wean ourselves from fossil fuels, and do so quickly, the *real* problems will begin after the middle of the twenty-first century. Sir John Houghton, one of the world's leading experts on global warming, told the London *Independent*: "We are getting almost to the point of irreversible meltdown, and will pass it soon if we are not careful."[26]

The due bill for our use of fossil fuels is now being served. By 2015, scientists had figured that "burning the currently attainable fossil fuel resources is sufficient to eliminate the [Antarctic] ice sheet."[27] This study was directed at Antarctica only, but all other ice would melt at the same time. How much time may be required to produce an ice-free planet? No one really knows. At present rates of increase, the actual burning of fossil-fuel reserves may take place within a thousand years. Complete melting of the ice, factoring in delays of thermal inertia, may require several thousand years. The momentum of this inertia would be irreversible, however.

Global warming is a deceptively backhanded crisis in which thermal inertia delivers results a half-century or more *after* our burning of fossil fuels provokes them. Our political and diplomatic debates react *after* we see results. Political inertia plus thermal inertia thus presents the human race and the planet we superintend with a challenge to fashion a new energy future *before* raw necessity—the hot wind in our faces—compels action. Global warming is dangerous because it is a sneaky, slow-motion emergency, demanding that we acknowledge a reality centuries in the future with a legal and diplomatic system that reacts in the past tense. Ken Caldeira, a researcher at Stanford University's Carnegie Institute of Science, one of the study's four co-authors, told Chelsea Harvey of *The Washington Post*, "The legacy of what we're doing over the next decades and the next centuries is really going to have a dramatic influence on this planet for many tens of thousands of years."[28]

In 2016, the atmospheric level of carbon dioxide breached 400 parts per million in all areas, at all seasons. Levels of methane and nitrous oxides, the two other principal greenhouse gases, also reached record levels by substantial margins. During 2015 and 2016, world temperatures, stoked by El Niño conditions, surged to a new record as well, above 2014's previous high. "We're moving into uncharted territory at a frightening speed," said World Meteorological Organization Secretary General Michel Jarraud.[29] Radiative forcing of these gases had increased 36 per cent since 1990.

During December 2015, January 2016, and February 2016 (meteorological winter), temperatures not only set world records, but did so by the largest margins (anomalies) since record-keeping began, in about 1880. Record temperatures continued, month by month, through at least July of 2016, the hottest month on record world-wide. February's global temperature was 1.35 degrees C above the 1951–1980 average, exceeding the previous record anomaly set in January, 1.13 degrees C, according to NASA's Goddard Institute for Space Studies (GISS). December 2015 was 1.11 degrees C above the same set of averages. Higher latitudes were the warmest: "Much of Alaska into western and central Canada, as well as eastern Europe, Scandinavia and much of Russia were at least 4 degrees Celsius (roughly 7 degrees Fahrenheit) above February averages, according to NASA/GISS," wrote Weather.com in February 2016.[30]

As radical rises in worldwide temperatures startled scientists early in 2016, James Hansen and 18 co-authors published a study in the open-access journal *Atmospheric Chemistry and Physics* (from the European Geophysical Union), making a case that several meters in sea-level rise could take place within a century, not the several hundred years projected by many scientists.[31] "My interpretation is that this is the beginning," Hansen said. "And it's one or two decades sooner than in our model."[32] "I think almost everybody who's really familiar with both paleo and modern is now very concerned that we are approaching, if we have not passed, the points at which we have locked in really big changes for young people and future generations," Hansen said.[33] "Limiting global temperature rise to 2 degrees C. (3.6 F.)" over pre-industrial levels, as recommended by recent diplomatic efforts such as the 2015 Paris accords, will not prevent climate-driven changes that will force evacuation of many coastal cities, Hansen and colleagues warned.

The science is evident on this matter, but the mother-lode mode of thinking still enjoys something of a lock on the political rhetoric of the U.S. Republican Party. Witness the party's 2016 nominee, Donald Trump, dismissing the entire body of evidence in two words ("Chinese hoax!"), as he argued for increased mining of coal to restore jobs. After taking office, Trump and his Environmental Protection Agency head, Scott Pruitt, issued orders opening many federal lands to increased coal mining and oil drilling.

Dawn of the Anthropocene

Increases in levels of greenhouse gases in Earth's atmosphere are part of a broader, intensifying, trend in a geological epoch now widely called the Anthropocene, in which human activities have been the primary force altering the planet "sufficiently to produce a stratigraphic signature in sediments and ice that is distinct from that of the Holocene epoch."[34] The Anthropocene actually began tens of thousands of years ago according to some, with the first use of fire and deforestation, but intensified with the advent of fossil fuels about 1800 C.E. By 1950, humanity's role in shaping of the Earth system was dominant, as population and industrialization exploded.

The controlling role of humanity includes not only infusion of carbon dioxide, methane, and other greenhouse gases into the atmosphere at levels previously not experienced since the Pliocene, 2 to 4 million years ago (with consequent increases in temperatures), but also rising levels of numerous artificial pesticides and herbicides, lead, fly ash and other forms of air pollution, fertilizers, plastics, and radioactivity resulting from nuclear weapons testing. Nuclear radioactivity from past tests will be detectable in ice and sediments for at least 100,000 years.

"Unlike with prior subdivisions of geological time, the potential utility of a formal Anthropocene reaches well beyond the geological community," commented Waters and colleagues in *Science* (2016).[35] "It also expresses the extent to which humanity is driving rapid and widespread changes to the Earth system that will variously persist and potentially intensify into the future."[36]

Several well-known climate scientists wrote in *Nature Climate Change in 2016* that "The next few decades offer a brief window of opportunity to minimize large-scale and potentially catastrophic climate change that will extend longer than the entire history of human civilization thus far,"[37] A team of 22 climate researchers, led by Peter Clark, from Oregon State University said:

> Most of the policy debate surrounding the actions needed to mitigate and adapt to anthropogenic climate change has been framed by observations of the past 150 years as well as climate and sea-level projections for the twenty-first century. The focus on this 250-year window, however, obscures some of the most profound problems associated with climate change. Here, we argue that the twentieth and twenty-first centuries, a period during which the overwhelming majority of human-caused carbon emissions are likely to occur, need to be placed into a long-term context that includes the past 20 millennia, when the last Ice Age ended and human civilization developed, and the next ten millennia, over which time the projected impacts of anthropogenic climate change will grow and persist. *This long-term perspective illustrates that policy decisions made in the next few years to decades will have profound impacts on global climate, ecosystems and human societies—not just for this century, but for the next ten millennia and beyond.*[38]

As temperatures rise, the combination of ocean water expanding as it warms and the melting of ice found on land is raising sea levels. Human beings have an affinity for the oceans, and a large proportion of us live on or near coastlines. Glance at a map of the world and point out the large cities that will be in peril as sea levels rise a few feet—Shanghai, Kolkata, London, New York City, Miami, and many others. Increase the proportion of greenhouse gases in the atmosphere and change its circulation patterns, expanding convection patterns that meteorologists call "Hadley Cells," which causes declines in rainfall over some areas, expanding deserts. Harvests fail, and people go hungry. We are adapted to the climate, as are all of Earth's flora and fauna. When the climate changes, everything changes.

A change in the balance of trace gases in the atmosphere requires a fundamental change in the ways that energy is obtained and used by everyone on Earth—the ways in which we transport ourselves, heat and cool our homes, and manufacture nearly everything we use in daily life. Scientists now anticipate that three-quarters of known fossil-fuel reserves must remain in the ground if warming is to be kept at a tolerable level—requiring companies that exploit these reserves to eventually write down their asset values to zero. The mother-lode model is failing in the marketplace as Peabody Coal, the biggest miner of that dirty fuel on Earth, goes bankrupt as wind power flourishes—because it costs less!

Even as wind and solar energy advance, however, politicians flush with oil-company cash deny that global warming exists. Many other people have come to realize that building a sustainable future is not a luxury. The mother-lode mentality does not anticipate conditions seven generations hence, as many Native paradigms instruct. The Native paradigm is more in line with science's understanding of thermal inertia, by which the atmosphere reacts to today's greenhouse-gas emissions a half-century from now on land, and more than that in the oceans, The behavior of thermal inertia requires that we anticipate the future, and not merely react to present conditions.

More people are realizing geophysical limits. Today we are witnessing an energy-system paradigm shift. For example (one of many), in 2016 for the first time, half of Iowa's electrical power generation came from wind. At the same time, technological change, as always, generates fear of unemployment. Paradoxically, such changes also always generate economic activity. A change in our basic energy paradigm during the twenty-first century will not cause the ruination of our economic base, as some deniers of climate change believe. Appreciation of sustainable models will enhance the economy.

The proportion of carbon dioxide in the atmosphere continues to rise worldwide, however, a trend that has not changed since the beginning of the industrial revolution. That level reached 400 parts per million in 2015, as high as it was in the Pliocene, 2 to 4 million years ago, when sea levels were about 50 feet and temperatures 4 to 6 degrees F. warmer. This is a key figure, and one that indicates how much change has yet to be experienced because of thermal inertia, within the next few centuries. This cake is already being baked. In terms of geologic time, the change is coming about remarkably quickly. Carbon dioxide is a trace gas, a tiny proportion of the atmosphere; at 400 parts per million, it comprises only one-tenth of one third of one per cent of the air. It is, however, a remarkably efficient retainer of heat.

The price of loading the air with carbon dioxide and methane continues to rise. The Intergovernmental Panel on Climate Change (IPCC) in 2014 projected that by 2100 rising sea levels and storm surges probably will swamp some of Asia's largest cities, among them Mumbai, Bangkok, Kolkata, Dhaka, Shanghai, Ho Chi Minh City and Rangoon; In Europe, London will be at risk, and New York City, New Orleans, Miami, and others will succumb to seas that rise in storm surges. The IPCC also projects that a rise in the atmospheric carbon dioxide level to 430–480 parts per million will raise acidity levels in the world's oceans. By 2100 it will reach a level that will imperil nearly everything that produces a calcium carbonate shell, from some phytoplankton (the basis of the maritime food web) to corals, many of which are also in trouble due to rising water temperatures. Carbon dioxide is being injected into the oceans far faster than nature can neutralize it. Seawater is usually slightly alkaline, at about pH 8.2. The pH of the oceans has fallen 0.1 during industrial times. The scale is logarithmic, so a 0.1 change means a 30 per cent increase in the concentration of hydrogen ions. Under a business-as-usual scenario, the pH may fall by roughly 0.5 by the year 2100.[39]

Even as surviving Indigenous peoples and their cultures are pulverized by the industrial machine—and even as the ecologically unsustainable nature of this ferocious juggernaut becomes more obvious—the colonizers (who flatter themselves with the descriptor "first world") continue to learn, absorb, and change, even in the act of conquest. The important question eventually may become whether these dominant forces can change fundamentally enough, and quickly enough, to avoid a climatic apocalypse.

James E. Hansen, long-time director of NASA's Goddard Institute for Space Studies, and the first person to discuss global warming in a scientific context observed natural limits as early as 1981.[40]

> The bottom line is this: business-as-usual, if it continues for even another decade will be disastrous for the planet. We can have a stable climate, clean air, and an unpolluted ocean. And clean energies yield good jobs. It is up to the public to make sure that we get onto a path that stabilizes climate and allows all the creatures of Creation to continue to thrive on this planet.[41]

The effects of climate change are not theoretical, and they are not speculative problems that can be handed off to future generations. Economic activity around the world, as well as the lives of animals and plants, are being affected today by rising temperatures. This is not merely a matter of a few degrees on the thermometer, but of alterations in an environment that sustains all of us.

Greenhouse gases have no morals, loyalty, nor party affiliation. Carbon dioxide is not having a debate with us. It merely retains heat. Thus, in 50 years, when our children are grandparents, the planetary emergency of which we are now tasting the first course, will be a dominant theme in everyone's life, unless we act now. Within a decade or two, thermal inertia will take off on its own, portending a hot, miserable future for coming generations.

Can a system predicated on growth adapt to a sustainable world in which having less "stuff" will be preferable? How can we adjust our desires to fit a new world in which more is not always better? Will our basic values change along with our energy sources?

A major—perhaps *the* major question facing an Earth and its human denizens in a time of worldwide environmental crisis is: can capitalism change its character? A sustainable environment can make good business. Witness the growth of alternative forms of energy. Can capitalism factor respect for the Earth that sustains us all into its calculus of development? If so, it may be a positive force in a new, sustainable world. If not—if it retains attributes of the cancer cell—then ultimately, our progeny will inherit an exhausted, poisoned world.

Can capitalism, with its appetite for pell-mell (and often environmentally destructive) growth, survive in a new world in which geophysical reality demands that we restrain our demands upon the Earth? Are we ready to operate with an accounting system that brings us all to account for the toll that our activities exact on the Earth and its atmosphere? Can we fashion a system in which polluting the atmospheric commons is defined as a criminal act for which sizable fines are levied and people serve time in prison? Such a system would re-define some present-day free choices (e.g. to trash the commons) as illegal acts. At that point, with Indigenous advice, we will be truly respecting mother Earth.

Because thermal inertia serves us with the impact of today's carbon-dioxide emissions 50 to 150 years in the future, the geophysical system *requires* that we heed Traditional Ecological Knowledge to anticipate the effects of our actions on the seventh generation. To do otherwise virtually guarantees that future generations will inherit a scorched, desolate world. The survival of human peoples, as well as the plants and animals upon which we depend, *requires* planning seven generations hence. This is *not* an optional luxury. This is basic survival behavior for a sustainable world—and it goes without saying, one very important example of how Indigenous ideas inform the thoughts and actions of everyone. Chief Sitting Bull of the Hunkpapa Lakota said a century and a half ago: "Let us put our minds together and see what life we can make for our children."[42]

Conclusion

Native wisdom (Traditional Ecological Knowledge, in this case) reflects what scientists refer to as the laws of nature, illustrated here by the behavior of thermal inertia. It's very important to understand that planning for the seventh generation reflects the same kind of thinking required to solve the global climate problem. The world must anticipate effects before they seem obvious. That is what ties the "mother Earth" philosophy together with climate science.

Regarding the famous passage in *Genesis* 1:28, the text of the Hebrew Bible (the most basic text for biblical scholars) does not justify absolute dominion of man over nature (as, unfortunately, it has been often interpreted since the 17th century); it emphasizes stewardship. See, for instance, the official documentation of the European Ecumenical Assembly in May 1989 (Basel, Switzerland): "We are to be stewards in God's world. Stewardship is not ownership. God the creator remains the sole owner, in the full sense of the term, of the entire creation."[43]

The keepers of religious doctrine have realized just how effective humankind has been at subduing the Earth, and just how damaging that subjugation has become. During the summer of 2015, Pope Francis, who has become well-known for directly tackling many controversial issues, made climate change a Vatican priority by issuing an encyclical (essentially a policy statement) detailing how the burdens of global warming worldwide fall disproportionately on the poor. Indigenous peoples around the world bear a disproportionate burden of environmental damage. In the United States today, Native peoples often live on ruined, exhausted land, suffering toxic consequences. Fully one-third of the Superfund sites declared by the United States Environmental Protection Agency are on Native American lands.

Pope Francis' encyclical was part of a broader campaign by the Pope to advocate protection of the Earth and all of creation. The pope prompted Catholic theologians to re-interpret *Genesis* to emphasize stewardship over subjugation. "We are the first generation that can end poverty, and the last generation that can avoid the worst impacts of climate change," said United Nations Secretary General Ban Ki-moon at an international symposium on climate change at the Pontifical Academy of Sciences April 28, 2015, one of the events leading up to the encyclical.[44]

Awareness of economic development's costs animates environmental advocacy "This, then, is the nemesis that modern Western man, together with his imitators … has brought upon himself by following the directive given in the first book of *Genesis*," wrote the great English historian Arnold Toynbee. "That directive has turned out to be bad advice, and we are beginning, wisely, to recoil from it."[45]

Let us ask the tough questions. When it comes to sustainability, what *really* works? What *really* matters? In the long-run, can a capitalistic system change its character to embrace standards of performance not predicated on growth; ones that improve the quality of life rather than sheer production?

Let us explore the ideas that will make tomorrow work.

Discussion Questions

Please describe the scientific principle of thermal inertia as discussed in this chapter. What are its implications for global warming?

How does the Haudenosaunee (Iroquois) idea of planning for the seventh generation (e.g. 140 years into the future) relate to the scientific principles of thermal inertia?

How do concepts that envision Earth as "mother" relate to sustainability in environmental studies?

Discuss the concept of the Anthropocene and its implications for sustainability.

Notes

1. A version of this chapter was previously published in *The Cultural and Literary Nationalism of Fourth World*, Acharya Nagarjuna University, Guntur, India, December, 2016.
2. George Cornell, "Native American Perceptions of the Environment," *Northeast Indian Quarterly* 7, no. 2 (1990): 3.
3. Jace Weaver, *Defending Mother Earth: Native American Perspectives on Environmental Justice* (Maryknoll, NY: Orbis Books, 1996), 10.
4. Roger Dunsmore, *Earth's Mind: Essays in Native Literature*, 1st ed. (Albuquerque, NM: University of New Mexico Press, 1997).
5. Ibid., 7.
6. Ibid., 15.
7. Ibid., 39.
8. Jordan Paper, "Through the Earth Darkly: The Female Spirit in Native American Religions," in *Religion in Native North America*, ed. Christopher Vecsey (Boise, ID: University of Idaho Press, 1990), 14.
9. Black Elk, *The Gift of the Sacred Pipe: Black Elk's Account of the Seven Rites of the Oglala Sioux*, ed. Joseph Epes Brown (Norman, OK: University of Oklahoma Press, 1967), 105.
10. Luther Standing Bear, *Land of the Spotted Eagle* (Lincoln, NE: University of Nebraska Press, 1978), 192, 194, 200.
11. Johnson Donald Hughes, *American Indian Ecology* (El Paso, TX: Texas Western Press, 1983), 80.

12. Standing Bear, *Land of the Spotted Eagle*, 196.

13. Ibid., 251.

14. Ibid., 38.

15. Ibid., 165.

16. Ibid., 166.

17. Virginia Irving Armstrong, *I Have Spoken: American History through the Voices of the Indians* (Athens, OH: Swallow Press, 1984), xi–xii.

18. Kurt Russo, *Review of Asserting Native Resilience: Pacific Rim Indigenous Nations Face the Climate Crisis* by Zoltán Grossman and Alan Parker, *American Indian Culture and Research Journal* 37, no. 2 (2013): 235.

19. Alan Parker and Zoltán Grossman, *Asserting Native Resilience: Pacific Rim Indigenous Nations Face the Climate Crisis* (Corvallis, OR: Oregon State University Press, 2012), 38.

20. Ibid., 45–46.

21. Russo, Review of *Asserting Native Resilience*, 236.

22. Ibid., 234.

23. Carol Berry, "Scholar Daniel Wildcat in Discussing the Environment: It's Relatives, Not 'Resources,'" *Indian Country Today Media Network*, 1 October 2011, accessed 2 November 2012,http://indiancountrytodaymedianetwork.com/gallery/photo/scholar-daniel-wildcat-in-discussing-the-environment%3A-it%E2%80%99s-relatives%2C-not-%E2%80%98resources%E2%80%99-56599.

24. Naomi Klein, *This Changes Everything: Capitalism and the Climate* (New York: Simon and Schuster, 2014), 183.

25. Robert Johnson, *Carbon Nation: Fossil Fuels in the Making of American Culture* (Lawrence, KS: University Press of Kansas, 2014), 14, 19, 39.

26. Geoffrey Lean, "Global Warming Will Redraw Map of the World," *The Independent*, 7 November 2004, 8, https://www.independent.co.uk/environment/global-warming-will-redraw-map-of-world-5350562.html.

27. Ricarda Winkelmann et al., "Combustion of Available Fossil Fuel Resources Sufficient to Eliminate the Antarctic Ice Sheet," *Science Advances* 1, no. 8 (2015), http://advances.sciencemag.org/content/1/8/e1500589.

28. Chelsea Harvey, "Scientists Confirm There's Enough Fossil Fuel on Earth to Entirely Melt Antarctica," *Washington Post*, 11 September 2015, accessed 3 May 2017, http://www.washingtonpost.com/news/energy-environment/wp/2015/09/11/scientists-confirm-theres-enough-fossil-fuel-on-earth-to-melt-all-of-antarctica/?wpmm=1&wpisrc=nl_headlines.

29. Joby Warrick, "Greenhouse Gases Hit New Milestone, Fueling Worries About Climate Change," *Washington Post*, 9 November 2015, Accessed online 17 March 2016, https://www.washingtonpost.com/national/health-science/greenhouse-gases-hit-new-milestone-fueling-worries-about-climate-change/2015/11/08/1d7c7ffc-8654-11e5-be39-0034bb576eee_story.html.

30. Jon Erdman, "February 2016 Was the Most Abnormally Warm Month Ever Recorded, NOAA and NASA Say," *Weather.com*, 17 March, accessed 16 July 2016, https://weather.com/news/climate/news/record-warmest-february-global-2016/.

31. James Hansen et al., "Ice Melt, Sea Level Rise and Superstorms: Evidence from Paleo-climate Data, Climate Modeling, and Modern Observations that a 2 Degrees C Global

Warming Could Be Dangerous," *Atmospheric Chemistry and Physics* 16, no. 6 (2016): 3761–812.

32. Justin Gillis, "Scientists Warn of Perilous Climate Shift within Decades, Not Centuries," *New York Times*, 22 March 2016, accessed online 1 May 2016, http://www.nytimes.com/2016/03/23/science/global-warming-sea-level-carbon-dioxide-emissions.html.

33. Chris Mooney, "We Had All Better Hope These Scientists Are Wrong About the Planet's Future," *The Washington Post*, 22 March 2016, accessed 1 November 2016, https://www.washingtonpost.com/news/energy-environment/wp/2016/03/22/we-had-all-better-hope-these-scientists-are-wrong-about-the-planets-future/?wpmm=1&wpisrc=nl_evening.

34. Colin N. Waters et al., "The Anthropocene Is Functionally and Stratigraphically Distinct from the Holocene," *Science* 351, no. 6269 (2016): 138–47.

35. Ibid.

36. Ibid.

37. Mooney, "We Had All Better Hope These Scientists Are Wrong About the Planet's Future."

38. Peter U. Clark et al., "Consequences of Twenty-First-Century Policy for Multi-Millennial Climate and Sea-Level Change," *Nature Climate Change* 6, no. 4 (2016): 360–69, (emphasis added).

39. Allison Bailey et al., "Seawater pH Predicted for the Year 2100 Affects the Metabolic Response to Feeding in Copepodites of the Arctic Copepod Calanus Glacialis," *PLOS*, 19 December 2016, accessed online January 17, 2017, https://doi.org/10.1371/journal.pone.0168735.

40. Hansen et al., "Ice Melt, Sea Level Rise and Superstorms," 3,761–812.

41. Bruce E. Johansen, "Global Warming, 'Thermal Inertia,' and Tomorrow's News," *Nebraska Report*, (October, 2007): 8–9.

42. David Archambault, "Taking a Stand at Standing Rock," *New York Times*, 24 August 2016, accessed January 2, 2017, http://www.nytimes.com/2016/08/25/opinion/taking-a-stand-at-standing-rock.html.

43. Philip C. L. Gray, "Christian Stewardship: What God Expects from Us," *Catholic Education.org.*, September, 2001, accessed 8 May 2016, http://www.catholiceducation.org/en/culture/environment/christian-stewardship-what-god-expects-from-us.html.

44. Elisabetta Povoledo, "Scientists and Religious Leaders Discuss Climate Change at Vatican," *New York Times*, 29 April 2015, accessed January 30, 2016, http://www.nytimes.com/2015/04/29/world/europe/scientists-and-religious-leaders-discuss-climate-change-at-vatican.html.

45. Arnold J. Toynbee, "The Genesis of Pollution," *New York Times*, 16 September 1973, 219.

Bibliography

Archambault, David. "Taking a Stand at Standing Rock." *New York Times*, 24 August 2016. Accessed January 2, 2017. http://www.nytimes.com/2016/08/25/opinion/taking-a-stand-at-standing-rock.html.

Armstrong, Virginia Irving. *I Have Spoken: American History Through the Voices of the Indians.* Athens, OH: Swallow Press, 1984.

Bailey, Allison, Claudia Halsband, Ella Guscelli, Elena Gorokhova, and Agneta Fransson. "Seawater pH Predicted for the Year 2100 Affects the Metabolic Response to Feeding in Copepodites of the Arctic Copepod Calanus Glacialis." *PLOS,* 19 December 2016. Accessed online 17 January 2017. https://doi.org/10.1371/journal.pone.0168735.

Berry, Carol. "Scholar Daniel Wildcat in Discussing the Environment: It's Relatives, Not 'Resources.'" *Indian Country Today Media Network,* 1 October 2011. Accessed November 2, 2012. http://indiancountrytodaymedianetwork.com/gallery/photo/scholar-daniel-wildcat-in-discussing-the-environment%3A-it%E2%80%99s-relatives%2C-not-%E2%80%98resources%E2%80%99-56599.

Clark, Peter U., Jeremy D. Shakun, Shaun A. Marcott, Alan C. Mix, Michael Eby, Scott Kulp, Anders Levermann, et al. "Consequences of Twenty-First-Century Policy for Multi-Millennial Climate and Sea-Level Change." *Nature Climate Change* 6, no. 4 (2016): 360–69

Cornell, George. "Native American Perceptions of the Environment." *Northeast Indian Quarterly* 7, no. 2 (1990): 3–13.

Dunsmore, Roger. *Earth's Mind: Essays in Native Literature* (1st ed.). Albuquerque, NM: University of New Mexico Press, 1997.

Elk, Black. *The Gift of the Sacred Pipe: Black Elk's Account of the Seven Rites of the Oglala Sioux.* Edited by Joseph Epes Brown. Norman, OK: University of Oklahoma Press, 1967.

Erdman, Jon. "February 2016 Was the Most Abnormally Warm Month Ever Recorded, NOAA and NASA Say." *Weather.com,* 17 March 2016. Accessed 16 July 2016. https://weather.com/news/climate/news/record-warmest-february-global-2016/.

Gillis, Justin. "Scientists Warn of Perilous Climate Shift within Decades, Not Centuries." *New York Times,* 22 March 2016. Accessed May 1, 2016. http://www.nytimes.com/2016/03/23/science/global-warming-sea-level-carbon-dioxide-emissions.html.

Gray, Philip C. L. "Christian Stewardship What God Expects from Us." *Catholic Education.org.,* 2001. Accessed May 8, 2016. https://www.catholiceducation.org/en/culture/environment/christian-stewardship-what-god-expects-from-us.html.

Hansen, James, Makiko Sato, Paul Hearty, Reto Ruedy, Maxwell Kelley, Valerie Masson-Delmotte, George Russell, et al. "Ice Melt, Sea Level Rise and Superstorms: Evidence from Paleoclimate Data, Climate Modeling, and Modern Observations that a 2 Degrees C Global Warming Could Be Dangerous." *Atmospheric Chemistry and Physics* 16, no. 6 (2016): 3761–812.

Harvey, Chelsea. "Scientists Confirm There's Enough Fossil Fuel on Earth to Entirely Melt Antarctica." *Washington Post,* 11 September 2015. Accessed 3 May 2017. http://www.washingtonpost.com/news/energy-environment/wp/2015/09/11/scientists-confirm-theres-enough-fossil-fuel-on-earth-to-melt-all-of-antarctica/?wpmm=1&wpisrc=nl_headlines.

Hughes, Johnson Donald. *American Indian Ecology.* El Paso, TX: Texas Western Press, 1983.

Johansen, Bruce E. "Global Warming, 'Thermal Inertia,' and Tomorrow's News." *Nebraska Report* (October, 2007).

Johnson, Robert. *Carbon Nation: Fossil Fuels in the Making of American Culture.* Lawrence, KS: University Press of Kansas, 2014.

Klein, Naomi. *This Changes Everything: Capitalism and the Climate.* New York, NY: Simon and Schuster, 2014.

Lean, Geoffrey. "Global Warming Will Redraw Map of the World." *The Independent,* 7 November 2004. https://www.independent.co.uk/environment/global-warming-will-redraw-map-of-world-5350562.html.

Mooney, Chris. "We Had All Better Hope These Scientists Are Wrong About the Planet's Future." *The Washington Post,* 22 March 2016. Accessed 1 November 2016. https://www.washingtonpost.com/news/energy-environment/wp/2016/03/22/we-had-all-better-hope-these-scientists-are-wrong-about-the-planets-future/?wpmm=1&wpisrc=nl_evening.

Paper, Jordan. "Through the Earth Darkly: The Female Spirit in Native American Religions." In *Religion in Native North America,* edited by Christopher Vecsey, 3–19. Boise, ID: University of Idaho Press, 1990.

Parker, Alan, and Zoltán Grossman. *Asserting Native Resilience: Pacific Rim Indigenous Nations Face the Climate Crisis.* Corvallis, OR: Oregon State University Press, 2012.

Povoledo, Elisabetta. "Scientists and Religious Leaders Discuss Climate Change at Vatican." *New York Times,* 29 April 2015. Accessed January 30, 2016. http://www.nytimes.com/2015/04/29/world/europe/scientists-and-religious-leaders-discuss-climate-change-at-vatican.html.

Russo, Kurt. *Review of Asserting Native Resilience: Pacific Rim Indigenous Nations Face the Climate Crisis,* by Zoltán Grossman and Alan Parker. *American Indian Culture and Research Journal* 37, no. 2 (2013): 234.

Standing Bear, Luther. *Land of the Spotted Eagle.* Lincoln, NE: University of Nebraska Press, 1978.

Toynbee, Arnold J. "The Genesis of Pollution." *New York Times,* 16 September 1973, 219. https://www.nytimes.com/1973/09/16/archives/the-genesis-of-pollution.html.

Warrick, Joby. "Greenhouse Gases Hit New Milestone, Fueling Worries About Climate Change." *Washington Post,* 9 November 2015. Accessed online 17 March 2016. https://www.washingtonpost.com/national/health-science/greenhouse-gases-hit-new-milestone-fueling-worries-about-climate-change/2015/11/08/1d7c7ffc-8654-11e5-be39-0034bb576eee_story.html.

Waters, Colin N., Jan Zalasiewicz, Colin Summerhayes, Anthony D. Barnosky, Clément Poirier, Agnieszka Gałuszka, Alejandro Cearreta, et al. "The Anthropocene Is Functionally and Stratigraphically Distinct from the Holocene." *Science* 351 no. 6269 (2016): 138–47.

Weaver, Jace. *Defending Mother Earth: Native American Perspectives on Environmental Justice.* Maryknoll, NY: Orbis Books, 1996.

Winkelmann, Ricarda, Anders Levermann, Andy Ridgwell, and Ken Caldeira. "Combustion of Available Fossil Fuel Resources Sufficient to Eliminate the Antarctic Ice Sheet." *Science Advances* 1, no. 8 (2015). Accessed 1 May 2008. http://advances.sciencemag.org/content/1/8/e1500589.

Ways of Being Human

CHAPTER EIGHT

Spiritual Relations, Moral Obligations and Existential Continuity

The Structure and Transmission of Tlingit Principles and Practices of Sustainable Wisdom

STEVE J. LANGDON

Acknowledgments

My name is Steve J. Langdon—my Tlingit name is *At Jishagoìoni Eìesh*, translated as "Father of the Tool, Implement"—given to me in marvelously Tlingit practice because it was the name of John Darrow, an elder of the Klawock *Gaanax'adi* clan, who would have generationally been my grandfather, a key relationship in Tlingit society. In 1934, Darrow provided anthropologist Ronald Olson with exquisite stories handed down to him about his people, the *Hinyaa* Tlingit of Klawock.[1] The information was recorded in Olson's handwritten notes, copies of which I obtained and returned to members of the *Gaanax'adi* clan in Klawock and thereby I "returned" as John Darrow to them.

It has been my extraordinary good fortune since 1973 to work with a remarkable group of Tlingit elders throughout southeast Alaska. I wish to pay my deepest respects to them and thank them for their graciousness and patience in taking time to mentor, teach and train me in the ways of the Tlingit. The information and perspective presented in this essay—limited in depth and scope as it is—could not have been created without the generosity of the following people who shared their understandings with me: Clara Peratrovitch (Klawock/*L'eeneidi*), Alicia and Theodore Roberts (Klawock/*Shunkweidi* and *Gaanax'adi*), Dennis Demmert (Klawock/*L'eeneidi*), David Katzeek (Chilkat/*Cangukedih*), Paul Marks (Chilkoot/

Luk'ax'adí) and Joe Hotch (Chilkat/*Kaagwaantaan. Gunalcheesh*). Thank you to each one and to many others for graciously teaching me their knowledge and wisdom.

Introduction

Several years ago, I participated in a conference at the Santa Fe Institute on "Conceptual Innovation and Major Transitions in Human Societies" at which I presented a paper entitled "Salmon People and Long-term Cultural Success on the Northwest Coast." The paper contended that the "conceptual innovation" of humans regarding and behaving toward salmon as "people" similar in all crucial respects to themselves, established the framework for relationships and practices that mirrored how Tlingit related to other humans. In thinking about the title and substance of this book it seemed to me that the key test of "Sustainable Wisdom" is long-term cultural success—a flourishing of people and their culture—accomplished by beliefs and practices that, among other things, sustained and perhaps enhanced key aspects of the environment they carefully shared with other living (and nonliving) forms. This "cultural flourishing" for the Tlingit is certainly demonstrated by their long-term presence in southeast Alaska (for at least 6,000 years based on current evidence but generally considered to be much longer by the Tlingit themselves).[2] When European explorers arrived in the 18th century, they found people who had constructed a rich, vibrant, and fulfilling culture built on an abundant and productive resource base that through hard work, inventive genius and careful calibrated utilization brought bounteous returns to the Tlingit. Given this background, four questions are posed here concerning the theme of "Sustainable Wisdom" and its applicability to Tlingit existence: (1) what is Tlingit sustainable wisdom? (2) where does it come from? (3) how does it inform and channel practice? (4) how is wisdom, and practice derived from that wisdom, sustained by being successfully transferred so that it continuously informs the actions of one generation after another of Tlingit persons?

This paper will endeavor to convey the "wisdom" the Tlingit constructed to sustain the existence they were a part of for thousands of years with special attention to their relationship with salmon, the most important resource and relationship cultivated lovingly and with great care in traditional Tlingit society. And yet, while it is still identifiable and can be seen in operation, Tlingit wisdom has certainly come under assault since the assumption of U.S. jurisdiction in the region in 1867.[3] Assaults upon the Tlingit way of life, concepts, and cultural practices have been mounted by religious, political (including military), economic, and educational institutions of the United States. As a result, maintenance of this wisdom is a challenge now being recognized by many in the Tlingit community as they work to continue their rich cultural heritage in the face of substantial impediments to merely acquiring salmon, customarily the most significant foodstuff for their livelihoods.

Existence Scapes: What Is It Possible to Think and Do?

In developing a framework to consider how Tlingit, as human beings, see, believe, and behave, a background for encountering all human experience and behavior is required. The intellectual ancestry of the attempts to comprehend and translate Indigenous spiritual and religious thought and practice in the Americas can be traced to the late 19th century. The initial construction of "animism" was advanced by the early evolutionary anthropologist E. B. Tylor. He labelled beliefs that all living or other things (such as rivers, winds, and mountains) each had spiritual essences. Tylor described this view as a "deep-lying doctrine of Spiritual Beings ... which so forcibly conduces to personification" and "can give consistent individual life to phenomena that our utmost stretch of fancy only avails to personify in conscious metaphor"[4] In Tylor's view, such beliefs constituted the most primitive and earliest forms of human spiritual/religious thought that over time evolved to monotheism. "Animistic" beliefs were simply thought of as superstition and all ritual systems of recognition, interaction, or gifting were seen as superstitious. The American scholar Franz Boas, while rejecting the evolutionist hierarchical view of "animist" beliefs, provided little beyond descriptive accounts of the spiritual beliefs, and did not attend to the manner in which those beliefs led to practice and motivated relational behaviors. The label "animism" was applied to Indigenous American religious and spiritual systems by most American anthropologists until the middle of the 20th century. Then influenced by the ethnographic revisioning initiated by Bronislaw Malinowski in his studies of the Trobriand Islanders of the South Pacific, the ethnographer A. I. Hallowell examined, in detail, the concepts of the Ojibwa and how they informed spiritual and religious behaviors. He offered a new account that underscored the ontology of beings as a comparative site of investigation and emphasized the Ojibwa view of the sentience and "personhood" of the spirits of the animals, birds, and fish, along with the essential communicative and reciprocal nature of interactions.[5]

Elsewhere in North America, Harvey Feit and Adrian Tanner's work with Canadian Cree built on Hallowell's seminal notions to investigate the relations between humans and the animals, birds, and fish with whom they lived.[6] These scholars recognized that much of the conceptual thought and behavior of the Indigenous groups was driven by concepts of insuring the survival and reproduction of the species upon which people depended. They discovered that often the behaviors were built on accurate empirical observations that gave rise to practices that protected habitat and insured that animal, bird, and fish populations were sufficiently numerous to successfully reproduce. Their research demonstrated that the Cree did not see humans as superior and believed that animals were co-equal with humans and stood in relations of interdependence co-occupying their shared environment with equal status as beings.

In other locations around the world, the work of scholars such as Bird-David, Viveros de Castro, Descola, and others expanded and revisited the concept of "animism" looking at the systems of thought and behavior as distinctly interactionist built on ontological assumptions offering a decidedly different view of natural forms and distinctions than found in Western thought.[7] As constructionists, they mobilized the notion that the alternative ontologies [nature of the characteristics of living forms as not universally nor objectively constructed as proposed by science] were the basis for how humans in specific cultures thought and behaved. The label "perspectivism" was advanced to characterize those scholars who focused on how members of Indigenous cultures sought to comprehend the vantage points of the natural beings with whom they shared existence.[8] Turner comments, "Animism and perspectivism thus take as their point of departure a reconception of the relation of nature and culture through an exploration of Indigenous conceptions of the common subjectivity of cultural and natural beings."[9]

Recently, the writings of Indigenous North American scholars on the nature of beliefs and practices have made significant contributions to the comprehension of Indigenous Native American understandings about the natural world. In particular, Greg Cajete, Raymond Pierotti, and Ronald Trosper have conveyed new insights about how Indigenous North American societies constructed and behaved in the world through principles of interdependence and reciprocity to produce societies that were resilient and sustainable.[10]

The early anthropological observers of the Tlingit, notably Swanton, Krause, Oberg, and Emmons, provided descriptions of Tlingit "animistic" spiritual beliefs and recorded oral traditions but never investigated actual practical relations by observing and engaging with Tlingit in their natural interactions.[11] Nor did they examine the links between the spiritual beliefs, behavioral actions with nature, and ritual practices of the Tlingit world. While eminent anthropologist Frederica de Laguna's initial ethnographic research and later editing of Emmons' manuscripts into publication followed staunchly in the Boasian descriptive mode of presenting Tlingit life according to Western ethnographic categories, by the end of her life she recognized that Tlingit indeed both recognized and engaged with salmon as if they were "persons" of fundamental similarity to humans with whom positive relations had to be established.[12]

Through encounters with Tlingit thought, perception and behavior, I have formulated the notion of *"existence scape"* to consider the manner in which humans position themselves in existence. The existence scape framework postulates that entities occupy a dynamic location where processes and interactions are ongoing in time and space. As a presence and participant in existence, humans receive and create a stream of sensations. Sensations are interpreted through both physiological and cognitive processes which can generate alternate interpretations of phenomena depending on the experiences of the human being, the contexts in which they were raised being most significant.

From these sensations, humans construct of meaning—the basis of much behavior—guided by key concepts and feelings they have learned. Some forms of meaning are located in consciousness but some, perhaps many, are subconscious; think of the generalized state of feeling "creepy" or having the hair stand up on the back of your neck. Other meaningful behavior, responses and initiations, arise through these mediated processes relying on concepts and understandings that are learned and are the basis for creative engagements—both social and environmental.

A quick note on language here: the nature of the language and how it is used can have a profound effect on the interpretation of sensations and accounts given about those events. Linguistic channeling is an important feature of the existence scape as it makes believable the entities, forces, and processes that are found in the existence scape. Two examples from the structure of the Tlingit language and its practice in Tlingit life are offered to provide insight into this proposition. As Cruikshank has pointed out:

> Tlingit language reflects subtle differences between ordinary and extraordinary, commonplace and mysterious, safe and dangerous ... the language is rich in verbs and emphasizes activity and motion, making no sharp distinction between animate and inanimate [as defined in Western thought]. Hence, mountains, glaciers, bodies of water, rocks and manufactured objects all have qualities of sentience.[13]

Further, these entities both attend to human activity as communication and communicate to humans in their own way. Joe Hotch conveys this two-way communicative flow in the following statement about salmon and Tlingit interaction with them:

> ... when they're jumping, we are supposed to say "Ey Ho"; you see a fish jump, "Ey Ho" [then] they know they're being appreciated so they keep jumping. And I guess

our people say it so they can know which way it's going. Just keep saying "Ey Ho," and that's the way they want to be talked to; the fish want to be appreciated.[14]

A further postulate of the existence scape is that humans have constructed cultural systems of beliefs and practices from concepts and perceptions through which they frame existence. Central to cultural frameworks are cosmological and ontological beliefs and principles. An existence scape comprises the realm of possible understandings, behaviors, and creative responses given a set of core cosmological and ontological principles constructed and modified overtime by members of a cultural/linguistic group.

Humans come to experience existence scapes as *habitus*: a system of embodied dispositions, tendencies that organize the ways in which individuals perceive the social world around them, channeling their actions and reactions.[15] Shared *habitus* represents the way group culture and personal history shape the body and the mind, and as a result, shape social action. Further, the predilections of the *habitus* are deeply embedded in the neurological structures of the brain as well as in the endocrinal fountains of the body. Descola points out that "although most members of any given community will find themselves unable to state explicitly the elementary principles of their cultural conventions, they nevertheless appear to conform their practice to a basic set of underlying patterns."[16] It is through this critical dimension of shared *habitus* that various members of a cultural group can give highly similar responses to the question "What is going on here?" A person from another cultural group would more than likely provide a markedly response to that same question.

Tlingit: People and Culture

At the time of sustained contact and now, the Tlingit were the primary occupants of the region referred to as southeast Alaska, the coastal strip lying seaward from the coastal range of mountains in northwestern North America (see Figure 8.1). Several Tlingit groups reside inland in British Columbia and the Yukon Territory of Canada. The environment in which the Tlingit live is characterized as rain forest dominated by large evergreen trees and innumerable rivers, streams and rivulets. Their adaptation is predominantly maritime and coastal—salmon are the most important food resource for all Tlingit. Likewise, the cedar and spruce of the rain forest supply the primary materials for creating objects necessary to maintain life: housing, heat, clothing, and transportation.

Tlingit organize themselves in various ways and through their names, their ancestry, and locations of social placement have a remarkable sense of not only their own position but also that of others they meet by determining identity markers.

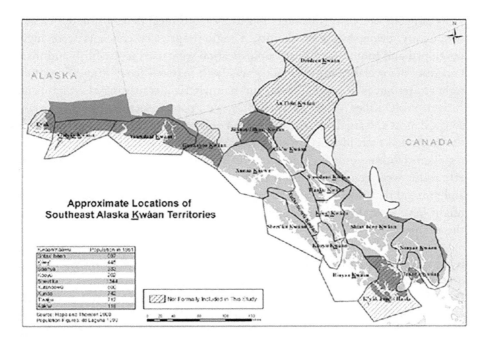

Figure 8.1: Tlingit Kwaan Territories.
Source: Author.

Geographically, Tlingit traditionally have been organized into about 15 *kwaans*, which are not tribes but instead are closely related clans, each of which have held separate territories in their kwaan region. Members of clans with recognized territories residing in the kwaan area typically intermarried and a state of truce prevailed internal to the kwaan among the clans. Rituals and ceremonies have played significant roles in maintaining social solidarity through formally constituted reciprocal acts of honoring and respecting clans and their members.

Socially, descent is traced through the mother's side in moieties (dual divisions—Raven and Wolf/Eagle), clans, houses and lineages. A crucial requirement of the social order, which resonates throughout cultural processes and into the *habitus* of each Tlingit, is what I have labeled "*obligatory reciprocity*."[17] A Tlingit person of a clan and moiety, cannot marry a member of their own clan or moiety. They must marry a person of the "opposite" side—thus a Raven moiety/clan member must marry a Wolf/Eagle moiety/clan member and vice versa. To do otherwise is considered incest and results in banishment and possibly death. Obligations that flow from this fundamental premise back and forth between houses, clans, and moieties are the means through which the society reproduces itself.

Politically, the clan (members linked by matrilineal descent), is the locus of sovereignty, ownership, and identity. Tlingit proprietary concepts were highly developed and protocols of great sophistication were used to establish and maintain respect for clan claims. Property was both material (consisting of territorial rights to resources in specific areas) and nonmaterial, (consisting of crests such as Beaver, Wolf, Killer Whale, Brown Bear, etc., names, sacred objects [*at.oow*],[18] songs, dances and regalia worn or displayed only on ceremonial occasions). Persons demonstrated their identity on such occasions, as well as in daily life, by wearing regalia or clothing on which an image of their crest was displayed. While property rights were generally respected, there were protocols for their establishment and renewal that were dutifully observed. If property rights were violated, Tlingit might have several recourses, including violence and seeking payments to rectify the wrongs.[19]

Tlingit demonstrate a rich and highly developed ceremonial system that incorporated aesthetic designs based on a well-established template and tradition of creating objects in wood, cloth, stone, bone, and antler. The use of these beautiful materials in ceremonies powerfully demonstrated critical elements of the belief system as well as offered the gift of aesthetic quality (beauty) to those who witnessed it. These performances restated and gave life to the beliefs through powerfully orchestrated mediums of display in rhetoric, dance, and song.

Core Conceptual Foundations of the Tlingit Existence Scape

The Tlingit existence scape was based on a number of conceptual foundations that linked beliefs, perceptions, emotions and behaviors together in a template of moral existence. These foundations include cosmological premises, essential relations among existential entities, obligations to conduct oneself in an appropriate manner in relations with existential entities, and through balance—a condition in which obligations to the interests and necessities of parties in an encounter are fully met.

A system of cosmological understandings about the nature of existence, its entities, forces, processes, and the nature of time and space is found in virtually all human cultures. Tlingit *cosmology* was grounded in the principle that all living entities had "spirits" and those entities cycle between domains of life in the world of direct experience followed after death by residence in another domain waiting for rebirth or reincarnation into this world. This process has been referred to as *cosmological cycling*.[20] The fact that spirits die and are reborn does not mean that time is cyclical—rather, it means that the past is distinct from the present; the future is only partially knowable (through the powers of the *ixt*—shaman)— and the entities and their relationship persist if each conducts themselves in the prescribed fashion. Despite the separation of the domain of the spirits from the domain of the living, movement between the domains can occur although such occurrences are unusual and highly significant.

Relations among entities were of great significance as interaction was unavoidable because entities were interdependent—their engagement and participation in appropriate fashion demonstrating respect and essential for existence. Interdependence means that relations not only existed but also were also inescapable and only through responsible actions would those critical connections be maintained. Cajete comments eloquently:

> ... people understood that all entities of nature – plants, animals, stones, trees, mountains, rivers, lakes, and a host of other living entities – embodies relationships that must be honored. Through the seeking, making, sharing, and celebrating of those natural relationships, they came to perceive themselves in a sea of interdependent relationships.[21]

Obligations derive from relations of interdependence that stipulate the moral dimensions of what entities, in particular humans, must carry out to insure the continuity of existence. These obligations are apparent in many aspects of Tlingit life. One of the most important is *respect* (*Ay wunei*). Respect is shown through interacting in a thoughtful, caring manner, carrying out key actions and speaking positively about another. Equally, if not more important, is the avoidance of disrespect. Disrespect is when one does not act appropriately or when one acts in an insulting manner toward the other being. Insulting acts include non-recognition, failure to conduct ritual action, and negative speech or behavior. Other obligations for moral behavior include engaging in appropriate ritual actions. These ritually prescribed behaviors are abundant and distributed through various daily practices as well as the activities of socially significant rituals.

Balance (*wooch yax*) is an absolute value that Tlingit strive to maintain because only when this is accomplished is order in place that allows positive functioning of elements. When it is not present, then actions must be identified and carried out by the parties involved, in order to attain it. Balance can be said to be operating when parties have conducted themselves appropriately and respectfully, and through their actions have met their obligations to treat others with respect and provide for their needs.

Tlingit beliefs, behaviors, and ritual practices are built on these basic principles and constitute a conceptual cosmological system built upon essential relations and articulated toward sustainability—*relational sustainability*.[22]

Shuka: translate this deep concept as "those born ahead of us who are now behind us and those unborn who await ahead of us"—thus the term references the past and the future.[23] It is a primary Tlingit concept that sits at the core of the existence scape and defines how Tlingit think of the social interactions among the generations that are essential for the continuity of the human spirits of the clan. It is the embodied Tlingit construct that culturally defines "cosmological cycling" through its direct connection to one's own relatives. The concept is always invoked at critical moments in Tlingit ceremonial and ritual events such as the mortuary potlatch or *koo'eex*. While primarily intended to recognize and honor recently deceased persons, this ceremony is a pivotal social moment in that it is formally structured to honor and recognize all the ancestors, those who have gone before, and to celebrate and embrace the positioning of young people in their social stations envisioning their active roles insuring relational sustainability into the future.

Tlingit Existence Scape: An Ontology of Willful Relational Interdependent Beings

Julie Cruikshank has observed that the Tlingit occupied "a moral universe inhabited by a community of beings in constant communication and exchange."[24] What is crucial to this community of beings is the relations that are established between them, thus *relationality* is an absolute characteristic and requirement of the Tlingit existence scape. Relationality through which beings recognize and orient to the existence of others is at the moral core and is the foundation of interdependence.[25] In the Tlingit existence scape, relationality permeates and is manifest in the cognitive constructs and orientations employed in daily life.

Relational cosmology is demonstrated in the mythic accounts of Raven (Yeil), the trickster transformer. Raven's interactions with others brought about numerous transformations in the basic conditions and processes of existence. Through these relational engagements with other entities, Raven unwittingly and unintentionally created the conditions of the world as currently experienced.

Relational ontology is manifest in the Salmon Boy story, the Tlingit mythic charter/covenant that establishes the essential "personhood" of salmon and the necessity of appropriate interdependent interactions. This account establishes the ontological status of salmon as "people" and frames relations between humans and salmon as members of the same ontological order who by dint of their being and their aim to continue being must engage in appropriate actions toward each other.

Relational epistemology is the premise that for Tlingit cognition, the acquisition of knowledge is the result of interaction with other entities. Personal experiences utilizing the concepts, perceptions, and emotions of the existence scape are essential for Tlingit as the basis for comprehending existence. This cognitive premise produces an orientation among Tlingit of careful attentiveness to one's experiences and highlights empirical interactions and engagements with other entities as locations for both reflective inquiry to obtain meaning and reflexive contemplation concerning the impact of one's own behaviors on other entities.

Relational phenomenology constitutes the basic pose, the set position for Tlingit existence in the world. All interactive engagement is necessarily relational and reflection and reflexivity are essential to interpret the meanings of such encounters. Interactions and engagements with other entities are always explored from the vantage point that actions of others may contain messages and can be understood to be volitional, directed acts of communication.

Relational sustainability is the set of principles through which Tlingit understand and engage existence. It is the set of beliefs and practices that orient Tlingit life and give meaning and purpose to all action. It is the recognition that the interdependent necessity of entities demands appropriate action by all to and for each other in order for existence to be continued. Cosmological cycling can only persist if appropriate relations are occur so that existence is sustained.

Sustainable Wisdom: Tlingit and Salmon

Several years ago when conducting research with Tlingit elders, I posed the question, "What were you taught about salmon as a child?" Almost to a person, men and women provided a version of what is known as the Salmon Boy story, also known as *Aakwatatseen* (his name), which translates as "Alive in the Eddy," or "Moldy End."[26] This story, or legend, is a mythic charter establishing for Tlingit the fundamental characteristics of salmon as people and conveying the manner in which humans are to respect salmon to insure their return in the future. It can also be considered a covenant as the account stipulates what human behaviors would make it possible for the salmon to return. The salmon boy story clearly and exquisitely conveys the basis for Tlingit attitudes and behaviors toward salmon that constitute *relational sustainability*.

Noted Tlingit scholar Frederica de Laguna explained how Tlingit thought of and behaved toward salmon as follows:

> The Tlingit ... felt that they were living in one world with the plants and animals and fish. The Tlingit thought of these too as like people with intelligence and moral values. They did not think that these were resources to be "managed." They felt that they should treat the fish and game and plants that they took with the respect that

one person would give another because they believed that the animals permitted human beings to use their bodies provided they treated them with respect and were not wasteful.[27]

The deeply reverential relationship between Tlingit and salmon is also demonstrable through its infusion found throughout Tlingit existence. Honored Tlingit elder and poet Nora Dauenhauer characterizes the relationship as follows:

> Not only have we always used salmon as our main diet, and not only has it been the mainstay of our subsistence and commercial economies, but the different varieties of salmon are a part of our social structure and ethnic identity as well. Beyond the physical use of salmon as food, salmon have symbolic and totemic value. Many clans have salmon as their crest. My clan is called *Lukaax.ádi* in Tlingit ... Our principal emblem is the sockeye or red salmon. The *L'uknax, adi* derives its name from another place of origin, related to the Tlingit word for coho salmon, which is their principal crest. The *L'eeneidí* clan has the dog salmon as its crest and the *Kwaashk'ikwaan* use the humpy or humpback salmon.[28]

The Tlingit utilization of the various species of salmon as clan crests as described above is much deeper than merely social marking to distinguish one clan from another. Clans demonstrate their relationship to the crests in various ways, most notably through artistic representations of the crests that appear on various pieces of clothing and especially on regalia, such as clan hats and blankets. Dauenhauer says the display of these regalia provide "... the education of the future stewards of a clan crest or tribal lands traditional ... informing the upcoming generation how to handle crests properly, when to handle them, who should handle them and how to talk about them."[29] These artistic representations are demonstrations of respect for the salmon and can interpreted as gifts to the salmon who are spiritually present when the crest regalia are presented. See Figure 8.2.

The basics of the *Aakwatatseen* ("Moldy Boy") mythic charter are as follows. A young boy is hungry and asks his mother for food. She directs him to the last piece of dried salmon. He picks up a piece, discovers it is moldy and throws it down in disgust. His mother reprimands him for his disrespect and he runs out of the house. The boy runs out of house down to the beach to check his sea gull snares and accidentally falls into the water and is in danger of drowning. The Salmon "Chief," having observed the boy's action and his predicament, sent his salmon relatives to get the boy and transport him to the Salmon people's offshore village on the bottom of the ocean. When they arrived at the village, the boy was surprised to see houses similar to those that humans lived in. When he entered the "Chief's" house, the boy was amazed to see the salmon take off their skins revealing their underlying status as persons like himself. Over the next year, the boy observed

Figure 8.2: *Lukaax.ádi* crest blanket displaying sockeye salmon crest associated with owner-ship of a specific salmon spawning location.
Source: Author.

how salmon people lived and discovered that their lives mirrored those of human beings; further he was trained by the salmon leader on how salmon must be treated by humans when they return to their stream homes in order for them to be able to return to their homes and be reborn. He described various respectful behaviors salmon appreciated as well as the critical ritual behavior of returning the salmon bones to the stream or ocean to insure the rebirth and return of the salmon.[30] The next year the salmon "chief" directs his people to prepare to return to their home streams. As they approach the boy's home stream he spots his father fishing and his mother processing the fish. He is directed by the salmon chief to present himself to his father which he does and is captured and handed over to his mother. As his mother began to cut the gills, she discovered a copper necklace and recognized it as the same one her son had been wearing when he disappeared. She showed her husband and both then sought out the shaman for instructions. The shaman told the parents to place the fish high on a shelf in the smokehouse and leave it there overnight. The next morning the parents awoke to find that the fish had been transformed back into their son. He then related all he had observed and learned, as well as what he had been taught by the Salmon "Chief" about the treatment of salmon. The teachings told how to establish respectful relations with salmon and included the critical ritual return of the bones of the first salmon captured and consumed by the group each year to the stream so that they would be transported to their village under the ocean. There, the bones would be reconstituted into salmon people once again and the cycle would continue. His teachings were adopted and practiced by his relatives and spread widely. He continued on in life to become an important shaman for his people. The salmon boy story is both charter and covenant by its identification that salmon will return to give themselves to those who treat them in the manner described in the story. As a covenant, it establishes an obligatory moral claim on humans to carry out their responsibilities or endanger the future of both people and salmon.

Sustaining Wisdom: Social Reproduction

Archeological evidence indicates that the Tlingit culture and practices that were observed and documented following sustained contact with Europeans and Americans in the late 18th century began appearing around 6–4000 years ago.[31] Over the centuries, a number of developments can be identified as the Tlingit cultural template emerged and was refined. For example, the intertidal fish traps constructed for capturing salmon are at least 4000 years old and at least three significant modifications to their operation occurred between 3000 years ago and contact with Europeans.[32] Those modifications demonstrate continuous exploration of methods for salmon capture, technologies, and behaviors that reveal an accumulation of knowledge and practice that arose from close attention to salmon and a continuing application of the principles derived from the mythic charter of relationship. The principle of "tidal pulse fishing," which utilized intertidal stone and stake weirs and traps to capture salmon only on the ebb tide allowing them to ascend to their spawning streams on each high tide, appears to have emerged around 1000 years ago.[33] The subtle and nurturing relationship between Tlingit and salmon, as I have demonstrated, emerged over thousands of years and made possible the continuity of Tlingit society and culture through sustaining abundant runs of salmon found in the streams of southeast Alaska by the early explorers, traders and ultimately the commercial exploiters of the late 19th century. How, over these generations, were these conditions of existence maintained? I have discussed the subtle and sophisticated intertwining of the cosmology, ontology, phenomenology and epistemology which generated the existence scape of Tlingit life. How, then, are these conceptual, orientational, and emotional states—taken together functioning as a part of the Tlingit *habitus*—transferred from one generation to the next?

I would like to turn now to discuss the topic of social reproduction of the complex of beliefs, attitudes, and practices in Tlingit individuals and groups that sustained Tlingit culture in the region. Throughout the world, it is clear that human societies actively seek to transmit and embed cultural beliefs, attitudes, and practices from generation to generation for various reasons. Among the reasons are: ensuring the health of the environment and relations with key relatives (species), the continuity of the culture, and critically in societies such as Tlingit, the return through rebirth of all the living forms, including the human. These practices, socialization, or enculturation, among hunter-gatherers have been the topic of scholarly research over many years. Cultures of the Northwest Coast of North America have long been recognized as distinctive, if not unique, among hunting and gathering societies. During my forty years of working among Tlingit and Haida people, I have had the opportunity to observe and be taught on a number of occasions the techniques and practices through which children and young people come to believe the validity of the cultural concepts, attend to and see the world around them in a culturally congruent manner, behave in appropriate ways, and commit to an identity that creates a powerful sense of belonging and pride of ancestry.

Much of human cultural practice is transferred informally, indirectly and unknowingly. A great deal is embedded neurologically through repetitive iteration below the level of attentive consciousness. Examples of this include eye movement, spatial orientation, interpersonal contact, and personal spacing. Narvaez has made the case that moral foundations of human behavior and morality are deeply positioned in our neurobiology and get established at an early age in ways similar to the examples above. With her colleagues, she has developed a sophisticated and integrated theoretical approach called "triune ethics meta-theory" to explore and examine the complex interactions among the biological, social interactional and cultural conceptual in the establishment of human moral orientation.[34] As "biosocial creatures [humans[are co-constructed by their social experiences, especially in early life when built-in maturational schedules of different body/brain systems" are coordinated through specific patterns of interaction and care.[35] She points out that "A child's natural inclinations for sociality and being a good member of society are supported" such that "virtue develops organically rather being imposed by adult rules" or pronouncements.[36] A further elaboration on the organicity of the development of a sense of morality and moral responsibility in children is that it arises from the continuous exposure to thought and behavior of those with whom the child most frequently interacts. At the same time, the child experiences those actions in the environmental context wherein human life is conducted and thus sees both the connections with other species, their qualities and the qualitative manner in which other human beings relate to the environment.

Language as a critical channel of cultural orientation is one of the means through which perceptions and sensations are linked through synapses to provide meaning but it has not received the attention it deserves in establishing cultural behavioral prescriptions and predilections, in particular toward morality. Languages vary in many ways. Cajete points out that "The verb-based nature of Native languages is ... connected to the Native cosmological assumption that we live in an interrelated living world in perpetual creative motion."[37] As such, Native languages "pay special attention to the way the event or place is in a perpetual state of motion."[38]

By contrast, the cultural psychologist Richard Nisbett has pointed, Indo-European language focus on the object as singular entity, in a static state generally.[39] Further, Indo-European languages need the distinction of dynamic to shift from the default cognitive state of stasis when talking about things. Finally, the language about relationships and the contingent nature of entities is woefully underdeveloped in Indo-European languages.[40]

As noted earlier Tlingit linguistic characteristics include features that well serve the requirements of the existence scape for transformation and dynamic movement between states, as well as for the development of awareness of conscious, attentive, volitional capabilities of the spiritual entities with whom they co-exist. While communicated as linguistic syntax, the salmon boy mythic charter's teaching about the essential personhood of the salmon people and the moral covenant found in the salmon boy story equip Tlingit with implicit and explicit cognitive tools for existence. Other aspects of the upbringing of Tlingit children contributed to the acquisition of the cognitive and emotional skills to embrace and practice the cultural traditions handed down to them.

The rearing of children in Tlingit society took place in a social context of immediate family and relatives. Within several months after birth, a naming ceremony was held in which the mother presented the child to her relatives and those of the father.[41] A name was given that typically was of a clan ancestor whose spirit was seen as returning from the land of the dead to this world through the child. Thus existentially positioned, the child would then be raised in atmosphere in which social relationships were constantly attended to and the social identity of the young person would demonstrated on numerous occasions and through numerous channels.

Children began being incorporated in the daily activities of living at about age four.[42] These activities would include food gathering and processing, assisting adults as directed and a variety of chores around the house. The child would be involved in movements by canoe primarily between various sites for food production beginning in April and continuing until September or October when the family would return to the home in the winter villages. While traveling, the child would learn stories about the various places that were visited as well as the technical knowledge about tides, weather, and various environmental phenomena that were encountered. An important part of the training of the child was learning the placenames that linked the group to their special locations as well as providing the lingua for sharing spatial orientation with others.[43]

The child would soon be immersed in the annual fall ceremonies of the *ku'eex* with its lavish regalia, formal rituals recognizing the ancestors and the opposites and dramatic performances of songs and stories of clan *at.oow*. As understanding grew of the concepts and beliefs on which these rituals were built, Tlingit youth gradually came to recognize what they were witnessing and participating in the emotionally powerful ceremonial events—at once both dramatic and instructive— that were a critical part of the development of social responsibility and moral accountability in the young person. Often it was not until adulthood that the full emotional resonance and commitment to the existence scape emerged.

It is important to draw attention to the critical role of being in natural settings, thereby experiencing the states of exposed existence and the crucial features of interdependence. Seeing bears eat salmon, sea lions consume herring, and wolves track down deer provide examples of the similarities in how humans and their co-occupants participated in the cycle of existence. These events were carefully attended to because from such events knowledge and understanding were acquired. Tlingit accounts of being in nature are replete with references to being taught by the various species, an example of which is an account that attributes the origins of Tlingit construction and use of intertidal stone traps to catch salmon to watching bears obtain salmon from natural intertidal pools. Cajete notes that "biophilia"— the idea that human beings have an instinctual understanding and need for affiliation with other things helps account for the manner in which Indigenous American groups think of themselves that "underlies the transfer of knowledge from one generation to the next."[44] I would extend the concept to include how intensive, nearly continuous immersion in natural processes engenders a *habitus* characterized by attentiveness, perceptiveness, knowledge, and a profound comfort with being in such places. John Muir, in 1880, made the following observation about a young "*hoona*" (Tlingit boy) traveling with his family of eight that camped near his party one night upon their arising the next morning:

> A little boy about six years old, with no other covering than a remnant of a shirt that hardly reached below his shoulders, was lying peacefully on his back ... despising

wind, and rain and fire. His brown, bulging abdomen all the more firmly bent on account of the curvature of the ground beneath him, heaving against the rainy sky, bare as a glacial dome, the rain running down from top of it all around and keeping it as wet as a bolder on the beach. He is up now, looking happy, and strong and fresh, with no clothes to dry, and no need of washing while this weather lasts.[45]

Pedagogy refers to formal mechanisms and practices of conveying culturally prescribed information and understandings to children. Tlingit pedagogical practices were varied some of which were highly structured as adults recognized the absolute necessity to existence for young people to carry on the culturally appropriate actions.[46] Tlingit storytelling was an exceptionally prominent practice characteristic of evenings at home, in camps, and when traveling by canoe. "The myths, legends, and stories that were told ... served to instruct the [children] in the history of their [clan] and in the deeds of their ancestors."[47] Grandparents were generally the tellers and dramatic language and body motions were used to heighten the tension and apprehension as devices. Tlingit adults had a clear sense of the cultural curriculum that was to be transmitted to the next generation and storytelling was one of the primary mechanisms to the accomplishment of these objectives. Repetition was central to successful conveyance, as it was understood that only through continuing exposure to basic understandings from various sources would the younger generation come to understand and commit to the cultural beliefs and practices that were essential to Tlingit existence. Raven stories were told again and again, at key times, in significant locations, and for various purposes. Sometimes they were told for laughter around the fire at night and other times they were told in order to convey specific lessons.

An interesting pedagogical feature of the practice was that the storyteller began but typically did not finish a story in a given evening. The following evening, he would ask the young listeners to tell him where he had quit. Some would be able to do it but others would not. He would then tell them all to think on it. The story would continue or be started again depending on the kinds of responses that were given. Through this technique the elders would be able to determine which of the children were best suited to be able to effectively and accurately relay the clan history in the ceremonial protocols which would be a critical feature of their adulthood that demanded oratorical skills and accuracy. Further, the repetitive teaching plus recitation were a critical method "very important in selecting and training the successor of a chief."[48]

Direct formal instruction in life skills, technologies, and knowledge about the land and resources was also important, especially on the cusp of transition to adulthood. The first menses was a critical turning point in the life of young women. The young girl was separated from her family, sometimes placed outside the house in a nearby structure due to concerns that her emerging state was extremely powerful and, because she had not yet learned how to manage it, it may be dangerous to others living in the house. At the appropriate time, "She is taken in hand by her mother- if motherless, by the nearest female relative – and put under special training for a period of from four to twelve months, the difference of time depending upon the parents' social circumstances."[49] During the period of seclusion, the young woman received instruction on what to do and injunctions on what not to do, especially during her menstrual period. She also received training in the arts of preparing regalia and was commissioned to create a special piece for a clan elder which she would gift upon her return to normal society.[50] At the conclusion of her training, a special ceremonial potlatch was held to honor her "coming out" at which time "a cape with hood attached and long fringe sewed to the front of the hood is made for her out of fine skins; this she wears—the fringe covering her face—for a number of days, or until she is used to the public."[51] A predominant theme of the training throughout was careful attention to one's environment and the development of restraint in thought, speech, emotion, and motion to insure that respectful behav- ior would always be followed.

Through accounts I have heard and observations of situations I have made over 40 years, it is clear that Tlingit adults were insistent on children paying atten- tion to the stories that were told them in order to insure that the lessons and messages contained in them would be absorbed. One man told me that his elders selected him to be the carrier of history and cultural knowledge for his house and clan. In order to be trained for that status, he was required at about age 10 to live for one year with each of his grandfathers in their clan or "tribal" house. His job was to help them, run errands, and get water and wood. He recalled receiving numerous directed lectures on various topics. On one memorable occasion, his grandfather emphatically said "Sit down, listen and remember." On that occasion, he recalls being told that he carried a "spear" with him always and that he should be extremely careful in how he used it. It was not until decades later, he told me, that he realized what was meant by the "spear"—it was your tongue, and the admonishment about being careful using it referenced spoken words. Two aspects of this exchange demonstrate critical elements of Tlingit cultural practice—first, to attend to, remember, and periodically consider elder advice and commentary and second, to always carefully reflect on speech prior to its delivery, especially when directly addressing, or regarding other important entities.

Structured Discovery

One of the distinctive Tlingit pedagogical practices was told to me in story form by Clara Peratrovitch, a remarkable woman of the *L'eeneidi* clan who lived in

Figure 8.3: *Hinyaa Tlingit* petroglyph showing Raven myth of owl obtaining fire. Originally located on Fish Egg Island, now in Klawock totem park.
Source: Author.

Figure 8.4: Owl's original long beak. Tlingit elder Clara Peratrovitch points to the location on the petroglyph where Owl's original long beak is depicted as well as the sections that were burned off as a result of holding the piece of fire in its beak.
Source: Author.

Klawock, Alaska. *Structured discovery* was designed by Tlingit parents to utilize children's curiosity to expose them to specific experiences and phenomena and then respond to children's questions with explanation and exegesis. The people of Klawock and those from other *Hinyaa* Tlingit villages traveled every spring to the island of *Shaanda* where they had camps. There they all made preparations for the arrival of the herring in April when they would collect the spawn being deposited. Behavioral excess was proscribed and a quiet expectant, attentive atmosphere permeated the camp in a demonstration of respect for the expected arrival of the schools of herring.[52] Tlingit Indigenous knowledge codified the sensitivity of herring to the conditions in the area where they expected to spawn and the people wished to insure that the best possible conditions so the herring would feel welcome and comfortable. Special protocols for welcoming and inviting the herring to spawn with song and dance were developed. Tree branches were carefully chosen and placed in a specific direction only after the herring had begun to spawn.[53] The territory around Klawock and *Shaanda* was the territory of the *Gáanax'ádi*, the oldest of the Raven clans. Raven was their primary crest and the stories of Raven's mischievous behaviors were on display around *Shaanda* on seven carved petroglyphs positioned in various out of the way locations. The glyphic images on these seven petroglyphs provided mnemonic guides to the myths of the Raven cycle with each petroglyph dedicated to one myth. One of the petroglyphs depicted the myth in which Raven cajoles Owl into flying out in the ocean to obtain fire (Figure 8.3). In the following account, Tlingit-honored culture bearer Clara Peratrovitch describes her experience as a child in regard to the petroglyph:

I was about four years old, when my mother took me to this location. Actually, she didn't put me right to the point. She just let me find it myself rather than take me right to it, because it really brought my attention to the petroglyph. And I was so excited I called my mother [At this point in her narrative, Clara changed from speaking in English to speaking in Tlingit] "Look at this rock. How come it's different?"

And my mother says, she said, "Years ago, the location of villagers, people that lived around, to mark their places and having to know that one day the young generation may come across. So they put their crest, their legend on the rock. They carved that on there. And so this is what this is," she says.

And I looked at it, and looked at it, and I asked her, "Well, what is this one here?" [She said], "I want you to look at it—all different sides. I'm going to point out what side, where the story starts first."

So she said, "This is the legend of the owl: When the owl first came about, it had a long beak. And the Raven, when time began, the Raven picked out the one bird that had a long beak. And he wanted the flame from the ocean—the "fireball" they called it."

And he asked the owl, "I want you to," he told the owl, "I'd like you to go out to the ocean and to the fireball and to bring a flame in."

•

And so the owl says, "Ok."

He says, (Raven says), "We are going to use that flame to light up at night, to have our meals cooked, to start a fire of our own to be used among our people."

And so the owl agreed.

And this part of the owl [pointing out feature of petroglyph {Figure 8.4}], is the first long beak. It is carved onto the side of the rock because the rock wasn't long enough and so the beak got put on the side of the rock.

And the owl went out.

These are his eyes, and his beak here [pointing out features of petroglyph]. This is part of his wing.

And when he got out there he took a flame, he bit a flame and he started flying back to the main land. And as he was traveling his beak got smaller. That's where it's [pointing to feature of petroglyph]—the first one is there. And as it burned, it got to this section here, burned, burned [rotating picture of petroglyph.]

And as he came closer to the mainland, his beak got smaller and so to this day [pointing to feature on petroglyph]—here's his eyes, his ears—and to this day, the owl has a short, stubby beak.

And the owl brought the flame into the mainland, gave it to Raven and Raven says, "Thank you. From now on you're going be free of anything that has come to you, you will not be harmed, you will be treated with respect, because you stepped forward for the world. You brought the flame for the world, to use."

This episode is an important lesson about Tlingit pedagogy, at least in this case as an example of *structured discovery*. The child is brought to the material (petroglyph) which, when encountered, stimulates immediate interest and curiosity. The child then asks questions about the object that prompt the adult, mother in this case, to respond by telling the child the story. The practice is exceptionally artful in mobilizing the inherent inquisitiveness of the child and using that interest to embed the account deep in the child's *habitus*. This event made such a strong impression on Mrs. Peratrovich that she was able to recall the incident and give a detailed account of it nearly 70 years after it occurred.

Moral Pedagogy

In *Aakwatatseen*, the Salmon Boy mythic charter, the sequence of events was brought on by *moral transgression*: the disrespect shown by the young boy to salmon by speaking badly of it and by throwing it down in disgust. An instance of moral transgression is both unfortunate and troubling but is often utilized by Tlingit adults as an opportunity for teaching morally correct behavior. In contemporary parlance, such occasions are regarded as a "teachable moment."

A number of examples of such occurrences were provided in interviews with Tlingit elders about how they learned about salmon when they were children. James Martinez, a Klawock elder, provided the following account:

> One time ... I was down at the creek and ... at Karheen ... I was throwing rocks at the fish .. and trying to kill them, you know ... and ... she saw me and ... called out to stop me and she called me up to the house you know, and she said, "You know, ... it's bad to do that ..." She said, "That's what we eat ... you know? That's what ... keeps us going ..."
>
> And I thought about it you know, and she finally told me a story, you know that ... the story was that, a young, some young Native boy was killing the fish and laughing and ... and throwing them around in the creek and ... and ... she said that she knew that his mother "told him that it was no good to do that ... some day they will come after you."
>
> I listened to that and got kind of interested, you know.
>
> And she says that, "... during that time, he was, went out someplace and ... he fell overboard ... had a ... he disappeared ... nobody could find him ... and ... it was years when ... the old man was down the river ... and he speared a fish ... and speared this one fish ... brought it home, it was big ... and ... he cleaned it and put it up on the rafter ... so nothing would get it ... and ... when they went to bed ... they were laying there ... and ... this fish started flopping around ... you know, and he said, I thought that fish was dead, and so he got up and got the fish ... and ... took it out and he was going to head it so it would die ... and while he was heading it the beads that that kid had around his neck when he disappeared, was in the inside of the skin of that fish. And he came in and told his wife ... and they took it out ... and laid it on a piece of red cedar ... and ... when they finally, morning finally came when they got out there ... that fish was no longer a fish but the boy."
>
> So, that's the story I heard, you know? Kind of made me think about what I was doing, you know? I didn't want that to happen to me so ... I quit killing fish ... or playing with them in the river ... and ... I started treating things with respect.

In this account, we see that the Salmon Boy myth was told to Martinez by his mother at a time of his *moral transgression*. Perhaps due to the juxtaposition of his own immediate behavior with that of the behavior of the boy in the story, the myth became so cognitively embedded that Martinez could recall the episode 70 years later.

A widely held moral obligation among northern hunters and gatherers was not to waste, and to utilize all the resource. The prohibition against waste was very strongly inculcated throughout the upbringing of children.[54] In the late 18th century, many Indigenous leaders registered profound concerns about the destruction of salmon and its impacts on Indigenous communities to federal fisheries officials as reported by Moser.[55] A number of Tlingit elders, like Jessie Dalton (Hoonah/*T'akdeintaan*) commented on their revulsion at the practices of the commercial salmon industry and their failure to process and utilize all the salmon they captured leaving thousands of fish to rot.

Thomas Mills provided the following account concerning an occasion when he and his brothers violated the principle of no wastage:

> One time my brothers and I went up the lake, up the river and lake with clubs and just clubbed maybe 200, 300 of those sockeyes. And being kids we were just going to leave them there. And father got a hold of what we did. And he took us all up there and we stayed up there and brought them all, all the fish down to the mouth of the river. We had to gut 'em all and pack 'em all in pack sacks. He wouldn't even—he was so mad at us—he wanted to teach us a lesson so he wouldn't let us put the fish in a skiff and take it over to the landing. We had to walk the trail and carry it all the way back down and come back up until all the fish was brought down. But that was a hard lesson on us but we learned not to go up there and just slaughter them, just for nothing. Because they are up there for a purpose too, to spawn and keep the resource coming. And we were just young and foolish and we didn't understand all that stuff. The way father did it to us, we thought we learned our lesson real quick and we never, ever repeated it. Least ways I haven't.

It is fascinating that moral transgression and its corrective is at the core of the Salmon Boy mythic charter/covenant and also situated so powerfully in the memories of living Tlingit and how they came to comprehend and embrace the meaning of the Salmon Boy story.

Conclusion

The Tlingit of southeast Alaska developed and lived in an existence scape based on the understanding of essential spiritual qualities of all entities with whom they lived. Those spiritual entities exhibited generally similar qualities to "human" persons with whom they engaged in reciprocal and interdependent relations. The Salmon Boy mythic charter/covenant is an example of how the Tlingit understood and structured their relations that were designed to insure the continuous return of salmon through time. The sustainable wisdom of their existence scape allowed the Tlingit to flourish over thousands of years.

- Sustainable wisdom is that wisdom necessary for sustaining a flourishing existence over an extended period of time.
- Sustainable wisdom structures essential principles in a manner that convincingly and powerfully identifies moral obligations and claims and demonstrates the consequences of failure to abide by them.
- Sustainable wisdom arises in part from experiences of "biophilia" that demonstrate the fundamental truths of the myths and legends through

immersion in and attention to, in the terms of the existence scape, the phenomena of existence.

- Sustainable wisdom that provides for a bountiful and flourishing life with abundant resources and rich cultural experiences can only be effective if the orientation, practices and emotions that characterize the existence scape are successfully transmitted generation after generation.
- Sustainable wisdom sits in stories and rituals transmitted and transferred through both powerful examples in art, regalia, song and dance, and carefully constructed pedagogy.
- Sustainable wisdom is made possible by immersion in a system of multichanneled, multimodal experiences that create and embed a *habitus* of perception, sensation, and action that are synchronized with moral consciousness and action.
- Sustainable wisdom must be seated in neurons and synapses that generate orientations of moral responsibility that make it deeply discomfiting to engage in moral outrage.

The Tlingit cognitive, emotional, behavioral, and expressive elements of their existence scape provided its bearers with the resilient wisdom to sustain a flourishing culture for thousands of years. They constructed and lived in an existence scape that was at once nurturing but demanding and developed a set of practices through which they were successful in transferring and committing generation after generation of their members to conduct themselves in accordance with the Tlingit principles of sustainable wisdom.

Discussion Questions

For the Tlingit, what does relational responsibility entail?

Describe Tlingit spirituality.

Briefly describe the "Salmon Boy" story. What is its importance to the Tlingit?

What lesson do Tlingit learn from the "Salmon Boy" story?

How does Tlingit wisdom differ from your culture's view of wisdom?

Notes

1. Ronald Olson, *Social Structure and Social Life of the Tlingit in Alaska*, Anthropological Records 26 (Berkeley, CA: University of California Press, 1967); Harvey Feit, "North

American Native Hunting and Management of Moose Populations," *Swedish Wildlife Research* Supplement 1 (1987): 25–42.

2. Kenneth M. Ames, *Peoples of the Northwest Coast: Their Archaeology and Prehistory*, ed. Herbert D. G. Maschner (London, UK: Thames & Hudson, 1999).

3. Thomas Thornton, *Haa Léelk'w Hás Aaní Saax'ú—Our Grandparents' Names on The Land* (Juneau, AK: Sealaska Heritage Institute, 2012).

4. Edward B. Tylor, *Primitive Culture Researches into the Development of Mythology, Philosophy, Religion, Art, and Custom* (London, UK: John Murray, 1871), 834, 260.

5. Alfred Irving Hallowell, "Ojibwa Ontology, Behavior and World View," in *Culture in History: Essays in Honor of Paul Radin*, ed. Stanley Diamond (New York, NY: Columbia University Press, 1960), 19–52.

6. Harvey Feit, "James Bay Cree Indian Management and Moral Considerations of Fur-Bearers," in *Native People and Renewable Resource*, 1986 Symposium of The Alberta, Society of Professional Biologists (1986); Feit, "North American Native Hunting and Management of Moose Populations "; Adrian Tanner, *Bringing Home Animals—Religious Ideology and Mode of Production of the Mistassini Cree Hunters* (New York, NY: St. Martin's Press, 1979).

7. Nurit Bird-David, "'Animism' Revisited: Personhood, Environment, and Relational Epistemology," *Current Anthropology* 40 (1999): S67–S91; Eduardo Viveiros de Castro, "Cosmological Deixis and Amerindian Perspectivism," *Journal of the Royal Anthropological Institute* 4, no. 3 (1998): 469–88; Philippe Descola, "Constructing Natures: Symbolic Ecology and Social Practice," in *Nature and Society: Anthropological Perspectives*, ed. Phillippe Descola and Gisli Palsson (London, UK: Routledge, 1996); Philippe Descola, *Beyond Nature and Culture*, ed. Janet Lloyd (Chicago, IL: University of Chicago Press, 2013).

8. Ernst Halbmayer, "Debating Animism, Perspectivism and the Construction of Ontologies," *Indiana* 29, Special Journal Issue (2012): 9–23; Bruno Latour, "Perspectivism: 'Type' or 'Bomb,'" *Anthropology Today* 25, no. 2 (2009): 1–2.

9. Terry S. Turner, "The Crisis of Late Structuralism: Perspectivism and Animism: Rethinking Culture, Nature, Spirit, and Bodiliness," *Journal of the Society for the Anthropology of Lowland South America* 7, no. 1 (2009): 11.

10. Gregory Cajete, *Native Science: Natural Laws of Interdependence* (Santa Fe, NM: Clear Light, 1999); Raymond Pierotti, *Indigenous Knowledge, Ecology, and Evolutionary* (New York, NY: Routledge, 2011); Ronald Trosper, *Resilience, Reciprocity and Ecological Economics: Northwest Coast Sustainability* (New York, NY: Routledge, 2009).

11. Kalervo Oberg, *The Social Economy of the Tlingit Indians*, Monograph 55, American Ethnological Society (Seattle, WA: University of Washington Press, 1973); George Emmons, *The Tlingit Indians* (Washington, D.C.: Smithsonian Institution Press, 1990); John R. Swanton, "Social Conditions, Beliefs, and Linguistic Relationships," in the *Twenty-sixth annual report of the Bureau of American Ethnology* (Washington, D.C.: Government Printing Office, 1908); Aurel Krause, *The Tlingit Indians*, trans. Erna Gunther (Seattle, WA: University of Washington Press, 1956).

12. Frederica De Laguna, *Under Mount Saint Elias: The History and Culture of the Yakutat Tlingit*, vol. 7, Smithsonian Contributions to Anthropology (Washington, D.C.: Smithsonian Institution Press, 1972); Emmons, *The Tlingit Indians*; Frederica De Laguna, Letter to Judy Ramos, 2001.

13. Julie Cruikshank, *Do Glaciers Listen? Local Knowledge, Colonial Encounters, and Social Imagination* (Seattle, WA: University of Washington Press, 2005), 142.

14. Thornton, *Haa Léelk'w Hás Aaní Saax'ú*, 50; Pierre Bourdieu, *Outline of a Theory of Practice* (New York, NY: Cambridge University Press, 1977).

15. Bourdieu, *Outline of a Theory of Practice.*

16. Descola, "Constructing Natures," 86.

17. Steve J. Langdon, *Natives of Alaska: Traditional Living in a Northern* (Anchorage, AK: Greatland Books, 2014).

18. *At.oow is a Tlingit concept and practice that can be thought of as "sacred property." At.oow relationships are held by clans and are based on iconic events in clan history and oral traditions that are represented in specific objects made by clan leaders to symbolize the claims that are presented only on extremely significant, formal occasions such as potlatches.*

19. Emmons, *The Tlingit Indians.*

20. Ann Fienup-Riordan, *The Nelson Island Eskimo: Social Structure and Ritual Distribution* (Anchorage, AK: Alaska Pacific University Press, 1983), 191.

21. Cajete, *Native Science*, 178.

22. Steve J. Langdon, "Sustaining a Relationship: Inquiry into a Logic of Engagement with Salmon among the Southern Tlingits," in *Perspectives on the Ecological Indian Native Americans and the Environment*, ed. Michael Harkin and David R. Lewis (Lincoln, NE: University of Nebraska Press, 2007), 233–273.

23. Nora Dauenhauer and Richard Dauenhauer, *Haa Tuwunáagu Yís, for Healing Our Spirit: Tlingit Oratory* (Seattle,WA: University of Washington Press, 1990), 19.

24. Julie Cruikshank, quotation as found on the back cover of *Haa Léelk'w Hás Aani Saax'ú: Our Grandparents' Names on The Land*, ed. Thomas F. Thornton (Seattle, WA: University of Washington Press, 2012).

25. Cajete, *Native Science*; Douglas L. Medin and Megan Bang, *Who's Asking? Native Science, Western Science, and Science Education* (Cambridge, MA: MIT Press, 2014).

26. Swanton, *"Social conditions, beliefs, and linguistic relationships of the Tlingit Indians."*

27. De Laguna, Letter to Judy Ramos.

28. Nora Dauenhauer, "Five Slices of Salmon," in *First Fish, First People: Salmon Tales of the North Pacific Rim*, ed. Judith Roche and Meg McHutchison (Seattle, WA: University of Washington Press, 1998), 101–102.

29. Dauenhauer, "Five Slices of Salmon," 102.

30. See quote from Joe Hotch, above.

31. Ames, *Peoples of the Northwest Coast: Their Archaeology and Prehistory.*

32. Steve J. Langdon, "Sustaining a Relationship."

33. Steve Langdon, "Tidal Pulse Fishing: Selective Traditional Tlingit Salmon Fishing Techniques on the West Coast of the Prince of Wales Archipelago," in *Traditional Ecological Knowledge and Natural Resource Management*, ed. Charlies Menzies (Lincoln, NE: University of Nebraska Press, 2006), 21–46; Langdon, "Sustaining a Relationship."

34. Darcia Narváez, *Neurobiology and the Development of Human Morality: Evolution, Culture, and Wisdom* (New York, NY: W. W. Norton, 2014); also her *Embodied Morality: Protectionism, Engagement and Imagination* (London, UK: Palgrave Pivot, 2016).

35. Narvaez, *Embodied Morality*, 7.

36. Ibid., 6.

37. Cajete, *Native Science*, 184.
38. Ibid., 184.
39. Richard E. Nisbett, *The Geography of Thought: How Asians and Westerners Think Differently—and Why* (New York, NY: Free Press, 2003).
40. Nisbett, *The Geography of Thought*.
41. Emmons, *The Tlingit Indians*, 237.
42. De Laguna, *Under Mount Saint Elias*, 512.
43. Thornton, *Haa Léelk'w Hás Aaní Saax'ú*.
44. Cajete, *Native Science*, 99.
45. John Muir, *Letters from Alaska*, ed. Robert Engberg and Bruce Merrell (Madison, WI: University of Wisconsin Press, 1993), 72–73.
46. De Laguna, *Under Mount Saint Elias*, 512–522.
47. Ibid., 514.
48. Ibid., 514.
49. Florence Shotridge, "The Life of a Chilkat Girl," *The Museum Journal* VIII, no. 2 (1917), 102.
50. Ibid., 102.
51. Ibid., 103.
52. Thornton, *Haa Léelk'w Hás Aaní Saax'ú*.
53. Ibid., 70.
54. De Laguna, *Under Mount Saint Elias*, 513.
55. Jefferson F. Moser, *The Salmon and Salmon Fisheries of Alaska. Report of the Operations of the United States Fish Commission Steamer Albatross for the Year Ending June 30, 1898* (Washington, D.C.: Government Printing Office, 1899).

References

Ames, Kenneth M. *Peoples of the Northwest Coast: Their Archaeology and Prehistory*. Edited by Herbert D. G. Maschner. London, UK: Thames & Hudson, 1999.

Bird-David, Nurit. "'Animism' Revisited: Personhood, Environment, and Relational Epistemology." *Current Anthropology* 40 (1999): S67–S91.

Bourdieu, Pierre. *Outline of a Theory of Practice*. New York, NY: Cambridge University Press, 1977.

Cajete, Gregory. *Native Science: Natural Laws of Interdependence*. Santa Fe, NM: Clear Light, 1999.

Cruikshank, Julie. *Do Glaciers Listen? Local Knowledge, Colonial Encounters, and Social Imagination*. Seattle, WA: University of Washington Press, 2005.

———. Quotation, as found on the back cover of *Haa Léelk'w Hás Aaní Saax'ú: Our Grandparents' Names on The Land*, edited by Thomas F. Thornton. Seattle, WA: University of Washington Press, 2012.

Dauenhauer, Nora. "Five Slices of Salmon." In *First Fish, First People: Salmon Tales of the North Pacific Rim*, edited by Judith Roche and Meg McHutchison, 100–21. Seattle, WA: University of Washington Press, 1998.

Dauenhauer, Nora, and Richard Dauenhauer. *Haa Tuwunáagu Yís, for Healing Our Spirit: Tlingit Oratory.* Seattle, WA: University of Washington Press, 1990.

De Laguna, Frederica. Letter to Judy Ramos. Copy in possession of the author, 2001.

———. *Under Mount Saint Elias: The History and Culture of the Yakutat Tlingit.* Smithsonian Contributions to Anthropology. Vol. 7: Washington, D.C.: Smithsonian Institution Press, 1972.

Descola, Philippe. *Beyond Nature and Culture.* Edited by Janet Lloyd. Chicago, IL: University of Chicago Press, 2013.

———. "Constructing Natures: Symbolic Ecology and Social Practice." In *Nature and Society: Anthropological Perspectives*, edited by Phillippe Descola and Gisli Palsson, 82–102. London, UK: Routledge, 1996.

Emmons, George. *The Tlingit Indians.* Washington, D.C.: Smithsonian Institution Press, 1990.

Feit, Harvey. "James Bay Cree Indian Management and Moral Considerations of Fur-Bearers." In *Native People and Renewable Resource Management: 1986 Symposium of The Alberta Society of Professional Biologists.* Edmonton: Alberta Society of Professional Biologists, 1986.

———. "North American Native Hunting and Management of Moose Populations." *Swedish Wildlife Research* Supplement 1 (1987): 25–42.

Fienup-Riordan, Ann. *The Nelson Island Eskimo: Social Structure and Ritual Distribution.* Anchorage, AK: Alaska Pacific University Press, 1983.

Halbmayer, Ernst. "Debating Animism, Perspectivism and the Construction of Ontologies." *Indiana* 29, Special Journal Issue (2012): 9–23.

Hallowell, Alfred Irving. "Ojibwa Ontology, Behavior and World View." In *Culture in History: Essays in Honor of Paul Radin*, edited by Stanley Diamond, 19–52. New York, NY: Columbia University Press, 1960.

Krause, Aurel. *The Tlingit Indians.* Translated by Erna Gunther. Seattle, WA: University of Washington Press, 1956.

Langdon, Steve J. *Natives of Alaska: Traditional Living in a Northern Land.* Anchorage, AK: Greatland Books, 2014.

———. "Sustaining a Relationship: Inquiry into a Logic of Engagement with Salmon among the Southern Tlingits." In *Perspectives on the Ecological Indian Native Americans and the Environment*, edited by Michael Harkin and David R. Lewis, 233–73: Lincoln, NE: University of Nebraska Press, 2007.

———. "Tidal Pulse Fishing: Selective Traditional Tlingit Salmon Fishing Techniques on the West Coast of the Prince of Wales Archipelago." In *Traditional Ecological Knowledge and Natural Resource Management*, edited by Charlies Menzies, 21–46. Lincoln, NE: University of Nebraska Press, 2006.

Latour, Bruno. "Perspectivism: 'Type' or 'Bomb.'" *Anthropology Today* 25, no. 2 (2009): 1–2.

Medin, Douglas L., and Megan Bang. *Who's Asking?: Native Science, Western Science, and Science Education.* Cambridge, MA: MIT Press, 2014.

Moser, Jefferson F. *The Salmon and Salmon Fisheries of Alaska. Report of the Operations of the United States Fish Commission Steamer Albatross for the Year Ending June 30, 1898.* Washington, D,C,: Government Printing Office, 1899.

Muir, John. *Letters from Alaska.* Edited by Robert Engberg and Bruce Merrell. Madison, WI: University of Wisconsin Press, 1993.

Narvaez, Darcia. *Embodied Morality: Protectionism, Engagement and Imagination.* London, UK: Palgrave Pivot, 2016.

———. *Neurobiology and the Development of Human Morality: Evolution, Culture, and Wisdom.* New York, NY: W. W. Norton, 2014.

Nisbett, Richard E. *The Geography of Thought: How Asians and Westerners Think Differently—and Why.* New York, NY: Free Press, 2003.

Oberg, Kalervo. *The Social Economy of the Tlingit Indians.* American Ethnological Society. Monograph 55. Seattle, WA: University of Washington Press, 1973.

Olson, Ronald. *Social Structure and Social Life of the Tlingit in Alaska.* Anthropological Records 26. Berkeley, CA: University of California Press, 1967.

Pierotti, Raymond. *Indigenous Knowledge, Ecology, and Evolutionary Biology.* New York, NY: Routledge, 2011.

Shotridge, Florence. "The Life of a Chilkat Girl." *The Museum Journal* 8, no. 2 (1917): 101–03.

Swanton, John R. "Social Conditions, beliefs, and linguistic relationships of the Tlingit Indians." In *Twenty-sixth annual report of the Bureau of American Ethnology,* 391–485. Washington, D.C.: Government Printing Office, 1908.

Tanner, Adrian. *Bringing Home Animals—Religious Ideology and Mode of Production of the Mistassini Cree Hunters.* New York, NY: St. Martin's Press, 1979.

Thornton, Thomas, ed. *Haa Léelk'w Hás Aaní Saax'ú—Our Grandparents' Names on The Land.* Juneau, AK: Sealaska Heritage Institute, 2012.

Trosper, Ronald. *Resilience, Reciprocity and Ecological Economics: Northwest Coast Sustainability.* New York, NY: Routledge, 2009.

Turner, Terry S. "The Crisis of Late Structuralism: Perspectivism and Animism: Rethinking Culture, Nature, Spirit, and Bodiliness." *Journal of the Society for the Anthropology of Lowland South America* 7, no. 1 (2009): 3–42.

Tylor, Edward B. *Primitive Culture Researches into the Development of Mythology, Philosophy, Religion, Art, and Custom.* London, UK: John Murray, 1871.

Viveiros de Castro, Eduardo. "Cosmological Deixis and Amerindian Perspectivism." *Journal of the Royal Anthropological Institute* 4, no. 3 (1998): 469–88.

CHAPTER NINE

Sustainable Wisdom and Truthfulness

An Indigenous Spiritual Perspective

FOUR ARROWS (WAHINKPE TOPA), AKA DON TRENT JACOBS

*Hindu prehistory implies a first age of "truth-speaking"; The second age is of "truth-seeking";
the third of "truth-declining" and the fourth (present day) of "very little truth remaining."*
—Thomas Cooper[1]

Truthful Communications as a Sacred Practice

Indigenous communication was inseparable from the spiritual sense of interconnectedness to all things visible and invisible. People understood that forms of it, including art, prayer, song, dance, and dialogue, produced vibrational frequencies that moved through time and space. A kind of telepathic interplay between humans and other life forms co-existed with verb-based languages that stemmed directly from the sacred places that Indigenous tribes inhabited.[2] The idea of intentionally speaking untruthfully was, by all accounts, unthinkable.

Then came the conquerors who had long since departed from original ways of being in the world. With their strategic lies, noun-based languages, and divide-and-conquer strategies, communication amidst the colonized eventually lost much of its sacredness. Art became a commodity. Prayer became selfish and subject to strict orthodoxy. Truth remained important in principle and in rhetoric via official rules, laws, and religious scriptures, to counter deceptive communication. Today there is "very little truth remaining."

This process started long ago and early religions may have tried to recover truthfulness, at least at first. For example, early Buddhist writings say "Without a commitment to truth there is no Buddhist path."[3] However, Buddhism in China around 250 BC was attempting to counter deceptions that contributed to civil wars that raged after the conquest of the Yangshaeo Indigenous culture. Early Chinese thinkers learned to be more concerned with the consequences of a belief and not truth per se.[4] Chris Frasure of the University of Hong Kong writes "Rather than assessing whether assertions are true, early Chinese thinkers supposedly evaluated whether utterances are pragmatically assertible, using the term "*ke*" (permissible, acceptable). Throughout the pre-Han literature, *ke* is regularly used to express assessments of conduct or policies as socially or ethically acceptable or permissible."[5] Early yoga traditions of India were mostly about salvation. "Yoga is an analysis of the dysfunctional nature of everyday perceptions and cognition which lies at the root of suffering. Once one comprehends the cause(s) of the problem, one can solve it through philosophical analysis combined with meditative practice."[6] (Today it seems we can substitute "alternative facts" for truth.)[7]

Truthfulness also was vital in Hindu Scripture as a way to help stop rule-breaking: "All our activities should be centered in Truth. Truth should be the very breath of our life. Without Truth it is impossible to observe any principles or rules in life."[8] Similarly, in the Christian New Testament scripture's references to truth had a more obvious hierarchical or preventive context. John 18:37, "Everyone that is of the truth hearest my voice." Ephesians 4:25: "Therefore, having put away falsehood, let each one of you speak the truth with his neighbor, for we are members one of another." In other words, the organized religions addressed truth in light of a world where lost truth was being sought.

Once American Indians recognized that the lies and broken treaties of the Europeans were not a mental illness per se, they also began to speak about the problem, giving testimony to the importance of truthfulness in hopes of preserving this vital aspect of Indigenous worldview. An English translation of something Chief Joseph of the Nez Perce, aka *Hinmatóowyalahtqit* (Thunder Rolling Down the Mountain) said conveys this: "Our fathers gave us many laws which they had learned from their fathers ... That we should never be the first to break a bargain. That it was a disgrace to tell a lie. That we should speak only the truth ... This I believe and all my people believe the same."[9]

Even the early American philosopher, Henry David Thoreau, realized that a loss of Indigenous wisdom had resulted in the death of truthfulness. He was a philosopher of Nature and its role in the affairs of mankind. His philosophy stems from his intense studies of the Indigenous perspective. His major publications stemmed from 2,800 handwritten pages in his "Indian Notebooks."[10] Although he had to make his special trips into "Nature" to commune, he understood that Indigenous Peoples saw truthfulness as a priority for healthful living. "Rather than love, than money, than fame, give me truth."[11]

Quaker author and educator, Parker J. Palmer also articulates truth in ways that seem oriented to the Indigenous perspective. In his quest for truthfulness, he compares his own soul with that of a wild animal as being "tough, resilient, resourceful, savvy and self-sufficient."[12] He has even specifically referred to the "deep knowing of the American Indian people as one of the most neglected and abused resources on the continent."[13] In *The Courage to Teach*, he shows his understanding of the original Indigenous understanding that Truth is a sacred aspect of a web of life beyond human societies:

> Authentic spirituality wants to open us to truth—whatever truth may be, wherever truth may take us. Such a spirituality does not dictate where we must go, but trusts that any path walked with integrity will take us to a place of knowledge. Such a spirituality encourages us to welcome diversity and conflict, to tolerate ambiguity, and to embrace paradox ...) The hallmark of a community of truth is in its claim that reality is a web of communal relationships, and we can know reality only by being in community with it.[14]

Indeed, what seems to make truthfulness so intrinsic to lifeways in Indigenous cultures relates to a focus on relational interconnectedness with all that is. As mentioned earlier, truth is in Indigenous thinking is about universal vibrational frequency. All communication is conceived as sacred and as being part of a function of service to the highest potential of life in harmony. We are created and co-created, not just influenced, by our thoughts, words, songs, dances and art. Truth is both about survival and communion. There is no separation between humans and other forms of life. In *A Time Before Deception*, Thomas W. Cooper explains this beautifully: "Living in communion meant unveiling the essential reality of what was already present. It is no surprise, then, that accounts flourish which describe Natives talking to the earth, or listening for the earth's heartbeat, to discover the world's rhythm and wishes."[15]

Thus, truthfulness is intrinsic to the concept of oneness. Living according to both requires observations of life's realities that seek complementarity, even with apparent opposites. Such observation requires setting aside preconceptions and ego attachment. Today this is difficult for most of us. We seem to have lost our ability to maintain consciousness long enough for this kind of observation. Indigenous wisdom teaches us that we can regain the ability if we become more sincere and more autonomous in our thinking. Under our dominant worldview, people are often slaves to external authority and to mechanistic process. We have forgotten that spirits also are partly responsible for what happens. We are also afraid of being too sincere. We are afraid of the suffering that can come from this. This fear dwells in our hearts and minds even before observation commences, before any object of fear can be recognized or articulated. Thus, from an Indigenous perspective, seeking truth calls for a circular path that starts with sincerity of mind and a subsequent courage to commit that triggers fearless engagement. After such engagement comes authentic reflection that leads to truthfulness and an enhanced sense of love of life that circles back to sincerity.

Black Elk speaks of such truth in terms of visions. Visions always come from the other-than-human world. When a vision comes from the thunder beings of the West, he says it comes with a terrifying storm and lightning. When one understands truth, one knows it is like lightning illuminating the dark clouds and that the dark and the light are both vital elements of it.[16] Recognizing this brings forth enlightenment and this in turn leads eventually to a greener, happier world. If we do not make sharp division lines between the lightning and the ominous clouds, between the self and the other, the truth is more likely to touch our hearts.

I wrote this reference to the Black Elk's storm metaphor on February 21, 2017 the day before the Water Protectors were to be forcefully evicted from the Lakota Treaty Lands at Standing Rock. The next day, just before the armed police made their move into the camp, I witnessed such a storm. The governor of North Dakota, putting forth the usual lies about reasons for the eviction had arranged for all sorts of local and state militia to conduct the raid. Around 2:00 PM, the sky over the main camp was clear except for corporate and police helicopters flying over. Surrounding the camp up in the hills and on Highway 1806 were hundreds of state troopers, local police, National Guardsmen, and Bureau of Indian Affairs police. Above them were black clouds and *lightning*. The contrast seemed so miraculous, a sign was put up with "truth" pointing to the camp and "fear" pointing to the hills. When the raid began, the clouds and sleet followed the raiders. Although the state and feds were able to accomplish their short-term mission and evict us, enough truth is continuing to emerge from the event to bring even more power to the larger movement to curtail the oil industries destruction of water systems. The loving courage and truth seeking of the Standing Rock ceremony is contagious and continues.

I talk a little more about the Standing Rock water protection movement later but want to share a personal and private ceremony in behalf of truth and fearless engagement. It is not a ceremony I recommend per se but I think it reveals connections similar to those experienced at Standing Rock. I do once or twice a week to keep me able to live in truth. Less than two kilometers from the beach in front of my palapa is an island with a cut into it and two caves that exit through the island to the west. Sometimes I swim to the island, but yesterday I paddled out on my stand-up board. I carried it into the grotto from the leeward side of the island, careful not to scrape it on the narrow walls while wading through the flow and ebb of the ocean. Inside, a gravel beach gives a few feet of dry land up against a vertical wall on the left. To the right, a more rugged wall rises 40 or more feet up into the island jungle. Through the wall are two tunnels that exit to the ocean on the windward side of the island. One is an underwater tunnel just wide enough for a human body.

On this day, the water was high and the in-and-out current was strong. I placed my board up against the inside wall. The sound of the waves crashing in and out through the smaller tunnel on the right was loud. For those unaccustomed to them, they might seem violent and frightening. I waded into the water to look out the exit hole. One large wave pounded through the opening, almost knocking me over. I held on to the rock wall and the exiting wave almost sucked me into the tunnel before I was ready.

I carefully studied the situation. Foolishness has no place in ceremony. I looked through the hole while waves pounded in and were then pulled back out. Daylight showed only for an instance between sets. I counted waves to see how bad it would be when the largest ones temporarily subsided. After several cycles, I determined it was doable if I went out right after the largest set came. I knew I would have to go fast before it returned. I launched myself into the cave, riding an outgoing flow until I was half-way out. Going as close to the bottom as I could and swimming hard while trying to avoid cutting myself on the narrow walls on either side of me. When the flow reversed, I managed to hold my place in spite of the power of the incoming wave. Knowing that the outgoing one would be strong enough to get me out into the opening, I waited. Then it came. Like superman I flew out into the daylight with only a few dolphin kicks. Knowing it was a matter of seconds before a wave could dangerously throw me against the rocks on its way back in, I swam with all my might to get into open waters.

When I returned to my board, I let out a loud scream to express my deep gratitude for being alive. I sat next to my board and meditated while listening to the roar of waves coming in stereo through the two openings. After a while there was complete silence. I opened my eyes. Remarkably, the crashing waves and rolling noise of the gravel had stopped for the first time. I looked around at the multi-colored rocks, shining from the water, in amazement. I noticed a rock crab scurrying across a boulder. I felt as if every cell in the jagged walls were my own. Or, perhaps I saw it as a brother, or a mother, or a father, or something greater than all of these combined. Tears came to my eyes. I gave thanks. Then as suddenly as it had stopped, the tumultuous action of the waves resumed. This time they seemed more a part of my soul than ever before. After praying, I went back out the way I originally came in.

The first thing I saw after exiting the grotto was a panga (type of boat) with two fishermen illegally pulling up a long net (I say illegally, because they were inside the boundary of the federally protected island). Normally when I see them I remind them that they are in violation of the law and are not giving the fish a chance to come back, or, sometimes I quietly curse them under my breath. This time, however, I felt I was them. I just waved and they waved back. I continued paddling home. Seagulls, pelicans, and clouds have long been my relations but now they were inseparable, a feeling I often have after the ceremony, but this day, perhaps because of the intensity of the waves, it was more.

Of course, one does not need to do such a radical ceremony as I have described above. I do this ceremony because on the Rio Urique in Mexico I had a near death event when I was accidentally sucked into a hole in which the entire Rio Urique disappeared.[17] It was life changing and the Spirits brought this other place into my life for a reason. But there are many fears in all of our lives we must face and many opportunities to seek truthfulness with fearlessness. Moreover, wherever we are located, we can find expressions of the natural world in which there are no options beyond that which is truth. They do not have to be as severe or as dramatic as my island cave for us to use them to remember who we are and why truthfulness is a vital pathway to help us do this.

Untruth as a Barrier to Remembering Indigenous Wisdom

Truthfulness has been shadowed by untruthfulness for a long time. One might think education might address this, but unfortunately it seems to merely rationalize the shadow. Although university presses like the one publishing this text have, on occasion, published books that reveal the truth about Indigenous realities, many have done the opposite. Unfortunately, the latter are the ones that make it into popular media. Such anti-Indian books by academic gatekeepers include the following:[18]

The Invented Indian (1990) by James Clifton, who writes that "acknowledging anything positive in the native past is an entirely wrongheaded proposition because no genuine Indian accomplishments have every really been substantiated" (p. 36);

Sick Societies: Challenging the Myth of Primitive Harmony (1992) by UCLA anthropologist, Robert Edgerton who writes about child abuse and other social maladies that were far more pervasive in primitive societies, proving the superiority of Western culture;

Lawrence H. Keeley's *War Before Civilization* that proposes that civilization and centralized governments have overcome the horrors of primitive life (2007);

Robert Whelan's Wild in the Woods: The Myth of the Peaceful Eco-Savage (1999) that offers such assertions as "Indigenous peoples have little to teach us about caring for the environment";

The Ecological Indian: Myth and History, by Shepard Krech, who asserts that the demise of the buffalo was the fault of the Indians themselves (2001).

Steven A. Lablank's *Constant Battles: The Myth of the Peaceful, Nobel* Savage that concludes that technology and science have put mankind on the right trajectory for world peace in comparison to the barbaric behaviors of aboriginal people (2003);

Steven Pinker's text, *The Better Angels of our Nature: Why Violence Has Declined* (2011) who uses exaggerated and erroneous stories about Indigenous violence against European colonists to make the case that we are better off now than in pre-state societies

Often popular historical novels, based on the kind of sloppy scholarship presented in the aforementioned texts, become best-selling mainstream books. Consider the recent New York Times best-selling "historical" novel by Bob Drury and Tom Clavin entitled *The Heart of Everything that Is (The Untold Story of Red Cloud)*. Newspaper book reviewers across the nation raved about the book's honesty. The high praise given it by such prestigious and progressive publications as *Salon* and the *Boston Globe* refer to the book's "exceptional fairness and accuracy." Yet, as I wrote in a critical book review entitled, "The Heart of Everything that Isn't," early on in their writing, after some gruesome descriptions about the savagery of Indian men, the authors assert incorrectly that the Lakota have always been a patriarchal society.[19]

I believe the hypnotic repetition of anti-Indian history (and here I also include romanticizing, as it is equally destructive) has prevented more support for Indigenous efforts to protect the last of the pristine landscapes on Earth and for recruiting local people with Traditional Ecological Knowledge in development projects. This is especially tragic when one understands that it is no coincidence that 80 percent of the biodiversity left on Mother Earth exists in the 20 percent of the land mass on which traditional Indigenous cultures still live.[20] Such cultures, still living according to the values that guided humanity for most of our time on this planet, are not as susceptible to the lies of neoliberalism as non-Indians for two reasons. This is why they are not only able to manage land sustainably but why they are also on the front lines around the world in trying to protect their waterways, as in Standing Rock.

This Indigenous truthfulness and the courage to continue honoring it continues to be sourced in knowing that everything is connected, as Thoreau learned from his own Indian studies. We can relearn to have such commitment to sacred communication by regularly communicating with the Earth and Water spirits with the kind of sincere observation skills I mentioned earlier. Sometimes entering into alternative modes of consciousness that changes our brain frequencies opens doors of perception in ways that reveal truths to us that cause us to challenge the hegemony all around. It is also important to reclaim a non-authoritarian perspective. This helps to prevent us from falling for the deceptive, persuasive language of strong personalities that may be trying to deceive us. Many good-hearted people who want to believe there is no climate change problem so they can avoid the fear have fallen for the linguistic strategies of Rush Limbaugh and Frank Luntz that essentially have convinced millions of people that our environment is fine. Now we have an entire administration who dismiss that human-generated climate change is a major challenge facing our world.

The same people who dismiss climate change dismiss Indigenous wisdom. Consider, for example, how Limbaugh talks about American Indian genocide in his 1993 book, *See I Told You So*. In these few sentences he uses at least four linguistic fallacies to get people to fall for his logic.

> In fact, while there were certainly atrocities against Indians by White people, there were just as many—and probably to a greater degree of savagery—committed by other Indians. Also, there are more American Indians alive today than there were when Columbus arrived or at any other time in history. Does that sound like a record of genocide. (p. 68)[21]

Sometimes academics truly believe what they are researching because they interpret data through the lens of a dominant worldview or bow to the writings of an "expert" in their field. When it comes to professional conclusions about truthfulness itself, Drs. Greg Cajete, Jon Lee and I found that many neuropsychologists rationalize lies as being a natural survival mechanism. In our text, *Critical Neurophilosophy and Indigenous Wisdom*, we counter this claim along with their studies that conclude that generosity exists only as a self-serving phenomenon.[22] Unfortunately, too few manage to counter such untruthfulness. Our early school experience does not really teach us independent, critical thinking, and false perspectives are inserted into textbooks by status-quo gatekeepers who make their way into school boards that select public school textbooks.

I remember a book published by McGraw Hill I used as an example of educational hegemony when I taught at Northern Arizona University. It was a required text for second graders in California. In one section, it asks the student what the California missionaries did for the Chumash Indians. The answer guide explained that they provided them with food and shelter. This is like saying the slave master's service to his slave was giving him a place to sleep at night. Another example of such educational hegemony in general is how Helen Keller is taught in most schools. Few readers of this chapter will know she was a socialist, a radical trade unionist, a suffragist, and an anti-war pacifist who courageously stood against the power structures of her day in spite of her personal challenges. Those in control of the destruction of life systems for the sake of money and power cannot allow us to have such heroes or heroines.

Because First Nations have such opposing views to the status quo, those in control of education manage to keep the truth at bay more often than one might imagine. The counter-balance to all of this is a return to Indigenous perspectives on truthfulness. Famous dissident and scholar Noam Chomsky declares: "The grim prognosis for life on this planet is the consequence of a few centuries of forgetting what traditional societies knew and the surviving ones still recognize."[23] I have talked to him and he agrees that the deep sense of truthfulness is largely what he meant. Fortunately, a number of people like the contributors to this book are working to bring such recognition to the larger world. In educational circles, authors like Jim Loewen (*Lies My Teacher Told me*), Howard Zinn (*A People's History*), Bill Bigalow (*Rethinking Columbus*) and my own texts (*Unlearning the Language of Conquest* and *Teaching Truly*) work to educate educators as well. Unfortunately, too few read such books and it will get worse it seems as I just learned that an Arkansas Republican just proposed a bill to prevent Zinn from being read in state financed schools.[24]

The Columbus Day Deception

Another untruth that is a barrier for learning Indigenous wisdom about truthfulness as a spiritual precept relates to a major holiday in the United States. It promulgates hurtful lies that continue genocidal and culturecidal treatment of the Indigenous and Mother Earth. When I was teaching social studies methods to teacher candidates at Northern Arizona University, during Columbus Day week, I gave an assignment for students to read some specific entries from the actual logbooks of Columbus and of some passages from the writings of Bartholomew de las Casas, a witness to the atrocities that followed his landing in "the new, undiscovered world." I did not tell them what to expect on the pages I sent them to discover. On the day the students were to share what they learned, two Navajo and two Hopi students came into my office and tearfully told me that in all their education in the state-controlled schools on their reservations they had never learned anything like what they learned about Columbus. Here are some of the readings I assigned. First are the ones written by Columbus himself:

In an April 1493, letter to Luis de Santangel (a patron who helped fund the first voyage), Columbus wrote:

> They are artless and generous with what they have, to such a degree as no one would believe but him who had seen it. Of anything they have, if it be asked for, they never say no, but do rather invite the person to accept it, and show as much lovingness as though they would give their hearts ... (nevertheless) their Highnesses may see that I shall give them as much gold as they need ... and slaves as many as they shall order to be shipped.[25]

And his March 4, 1493 letter to King Ferdinand of Spain:

> They are very marvelously timorous. They have no other arms than spears made of canes, cut in seeding time, to the ends of which they fix a small sharpened stick ... They are so guileless and so generous with all that they possess, that no one would believe it who has not seen it. They refuse nothing that they possess, if it be asked of them; on the contrary, they invite any one to share it and display as much love as if they would give their hearts ... They do not hold any creed nor are they idolaters; but they all believe that power and good are in the heavens ... In conclusion, to speak only of what has been accomplished on this voyage, which was so hasty, their Highnesses can see that I will give them as much gold as they may need.[26]

The second set of original writings I asked my NAU students to read were from the writings of Bartolomé de Las Casas (1474–1566). He might be thought of as one of the thirty-nine reported police officers from nine states who refused to join the militarized force against the Standing Rock water protectors. Some even resigned from their jobs.[27] Las Casas arrived as one of the first Europeans in the Americas, initially participating and then stridently opposing the atrocities started by Columbus. He gave up his own Indian slaves and courageously protested against what was happening. As a Dominican Friar in 1542 Las Casas wrote, *A Brief Account of the Destruction of the Indies*, motivated by his fear Spain would be punished by God and owing to his concerns for the souls of the Natives. He writes about how Columbus held the position of Viceroy and governor of the Americas from 1493 until 1500 during which time he initiated and institutionalized the systematic extermination of the Taino People. [28] His policies continued with the blessings of the Catholic Church and through the actions of the conquerors that followed. Las Casas writes:

> I also affirm that I saw with these Eyes of mind the Spaniards for no other reason, but only to gratify their bloody mindedness, cut off the Hands, Noses and Ears, both of the Indians and the Indianesses, and that in so many places and parts, that it would be too prolix and tedious to relate them. Nay, I have seen the Spaniards let loose the Dogs upon the Indians to bait and tear them in pieces, and such a Number of Villages burnt by them as cannot will be discovere'd: Farther this is a certain Truth, that they snatched Babes from the Mothers Embraces, and taking hold of their Arms threw them away as far as they would from them (a pretty kind of barr-tassing Recreation). They committed many other Cruelities, which shook me with Terror at the very sight of them, and would take up too much time in the Relationship. [29]

The students also read some passages like those these from the Las Casas eye-witness reports:

> They spared no one, erecting especially wide gibbets on which they could string their victims up with their feet just off the ground and then burn them alive. (p. 16)

> Yet another member of the governor's party galloped about cutting the legs off all the children as they lay sprawling on the ground. (p. 22)

> Indeed they invented so many new methods of murder that it would be quite impossible to set them all down on paper. (p. 23)

> Not a single Native of the Island committed a capital offense, as defined in law, against the Spanish while all this time the Natives themselves were being savaged and murdered. (p. 23)

> Both women and men were given only wild grass to eat and other unnutritious foodstuffs. The mothers of young children promptly saw their milk dry up. (p. 24)

> During the three or four months I was there, more than seven thousand children died of hunger, after their parents had been shipped off to the mines. (p. 30)

> A Spaniard who was out hunting deer or rabbits realized that his dogs were hungry and not finding anything they could hunt, took a little boy from his mother, cut his arms and legs into chunks with his knife and distributed them among his dogs. (p. 74) [30]

Although more and more people are learning the truths about our early history in the Americas, a larger percentage remains clueless or essentially convinced by revisionist histories that they watch on television. Granted, not all of us have access to truer descriptions or have the availability of time to triangulate primary source documents and to sift through controversial or contradictory claims. However, when truthfulness returns as a vital priority, and fearlessness returns as a way to move forward in the world, we generally can feel untruthfulness. This is more likely when our cognitive dissonance is still working. And if it is, the internet still offers opportunities to discover alternative histories and we can use our reasoning and intuitive minds to determine which truths are probable and how to change our views accordingly.

Standing Rock as an Example

Indigenous perspectives on truthfulness as a sincere respect for place and a way to experience oneness leads to "love that flows like a river."[31] What happened at Standing Rock has been and remains an example of this perspective in action. I made several trips to the main camp on the Standing Rock Reservation to participate in the water protection movement led by the First Nations of North America in an effort to require an Environmental Impact Statement for the Dakota Access pipeline, which we won under the Obama administration and lost illegally under President Trump. Once I served as a medic on the front line. The other times I served in my capacity as a Veteran for Peace member (and co-founder of Chapter 100) in helping teach arrestable water protectors the importance of maintaining peaceful, prayerful attitudes on the front lines. I wrote a number of articles for *Truthout* about the water protection movement there.[32] Each article describes both some particular event and how the Indigenous Peoples and their allies responded in accordance with traditional Indigenous wisdom about truthfulness.

I bring the Standing Rock Water Protection Movement into this chapter because I believe it exemplifies the contrast between a commitment to truthfulness practiced by most of the Water Protectors and submission to untruthfulness practiced by those in opposition to them. When the Indians from the many tribes at Standing Rock talked about truth, conversations were mostly about love, generosity, the river spirits, gaining consensus of understanding amidst disagreements, and allowing for unknowing. That it was about truthfulness, however, is unmistakable. Without the "either-or" context and in light of a sincere effort to find complementarity-even with the Dakota Access Pipeline employees and police—truth-seeking and truth-speaking is indistinguishable from oneness with all. Generosity and courage are constant. Fearless engagement was common. Prayers rather than anger was the response to the most egregious actions against them.

This does not mean there is acceptance of the lies from the state, federal agencies, or the media. It is quite to the contrary. Although the propaganda—about violence coming from the water protectors, about the process, about the terrorism practiced by the state, federal, and BIA troops, about the law, and about treaty rights—has been overwhelming, and although lies are the greatest barrier to a rebalancing of life systems, yet, people are realizing that after 500 years of such lies, a global awakening to Truth is happening. I was at Standing Rock when Donald Trump took office. A number of non-Indian allies were cursing or crying. The Indians around me sort of winked and one woman hugged a young lady who was in disbelief. "Welcome to our world honey," she said. "We've been dealing with this sort of thing for 500 years!"

Such humor seems to be a part of how Indigenous people hold onto the truth against all odds. Getting close to my word limit for this chapter, I wanted to make sure I did not miss something important and called a friend of mine who is a Cree Medicine Man and conducts Star Lodges and Sun Dances. I asked him if he had any words about the importance of truth and the problem of lying.

He said, "Well, we thought the Europeans had a mental disease and could not describe reality at first and now we pretty much know this is still true. Have you told any good jokes about George Washington?"

My reply was something like, "Uh, no. What do you mean?"

"Well, I was taught in Boarding School that he is the model for truth telling in the U.S. Something about a cherry tree," he said.

I replied, "Oh, yeah. He chopped it down and told the truth to his father I think and didn't get in trouble because he didn't lie. Something like that … but …"

He interrupted, "I knocked over an outhouse once and told my father that I did it when he asked and got whipped good. Course, maybe that was 'cause he was in it when I pushed I over."

What is happening in our world today are unnecessary, untruthful communications, beliefs and actions that are potential extinction events. We must all quickly learn to return to the Indigenous perspective on Truthfulness I have introduced here with love, compassion, determination, and humor to bring it to bear on transformations for the benefit of future generations.

We are all related.

Four Arrows (Wahinkpe Topa), aka Don Trent Jacobs

Discussion Questions

What is the relationship between fear, perception of authority, and untruthfulness in our dominant culture and how do traditional Indigenous perspectives differ?

Study the phenomenon of hypnosis via Google and consider the connections between untruthfulness, fear. and authority in terms of it.

How did the author make connections between habits of truthfulness and exposure to the natural world? What ways can you think of to use immersion in nature as a way to see what is true and what is not?

How is the continuing celebration of Columbus Day an example of how untruthfulness is supported in education? Why is the way Helen Keller is studied in schools an example of educational hegemony? (Google "The truth about Helen Keller").

Parker Palmer writes (cited above) that "The hallmark of a community of truth is in its claim that reality is a web of communal relationships and we can know reality only by being in community with it." Consider and discuss how this relates to traditional Indigenous communities and how it might apply in your own community (or not) in light of his definition of truth as "an eternal conversation about things that matter conducted with passion and discipline."

Notes

1. Thomas W. Cooper, *A Time Before Deception: Truth in Communication, Culture and Ethics* (Sante Fe, NM: Clear Light, 1998), 29.
2. Four Arrows, *Point of Departure: Returning to Our More Authentic Worldview for Education and Survival* (Charlotte, NC: Information Age, 2016).
3. Gil Fronsdal, "The Perfection of Truth." *Insight Meditation Center*, accessed 15 July 2018 http://www.insightmeditationcenter.org/books-articles/articles/the-perfection-of-truth/.
4. Donald Munro, *The Concept of Man in Early China* (Stanford, CA: Stanford University Press, 1969), 55.
5. Chris Frasure, "Truth in Pre-Han Thought" (2016), accessed 10 July 2018, http://cjfrase. net/site/uploads//2016/10/Truth_Fraser_web.pdf.
6. David Gordon White, "Yoga, Brief History of an Idea," accessed 22 June 2018, https:// press.princeton.edu/chapters/i9565.pdf.
7. Kellyanne Conway, "La La Land Blunder," *The Guardian*, 3 March 2017, accessed 01 August 2018, https://www.theguardian.com/us-news/2017/mar/03/kellyanne-conway-alternative-facts-mistake-oscars.
8. Mohandas Gandhi, "Gandhi on the Meaning of Truth" from *Yeravda Mandir*, chapter one, (1927), accessed online 28 April 2018, https://berkleycenter.georgetown.edu/quotes/ mohandas-gandhi-on-the-meaning-of-truth.
9. Chief Joseph, "On a Visit to Washington, D.C." (1879), accessed 04 April 2018, http:// www.firstpeople.us/FP-Html-Wisdom/ChiefJoseph.html.

10. Henry David Thoreau, *Indian Notebooks* (1847–1861), transcribed by Richard F. Fleck (2007), accessed online 17 April 2018, https://www.walden.org/wp-content/uploads/2016/03/IndianNotebooks-1.pdf.

11. Henry David Thoreau, *A Week on the Concord and Merrimack Rivers* (1849), accessed online 05 March 2018 https://archive.org/stream/weekonconcordmer1849thor/weekonconcordmer1849thor_djvu.txt.

12. Parker J. Palmer, *A Hidden Wholeness: A Journey to an Undivided Life* (San Francisco, CA: Jossey-Bass, 1999).

13. Parker J. Palmer, Quotation located on the back cover of *Teaching Virtues: Building Character Across the Curriculum* by Four Arrows and Jessica Jacobs (New York, NY: R&L Education: Scarecrow, 1981).

14. Parker J. Palmer, *The Courage to Teach: Exploring the Inner Landscape of a Teacher's Life*, 10th anniversary edition (San Francisco, CA: Jossey-Bass, 2007), ix, 51.

15. Thomas W. Cooper, *A Time Before Deception*, 168.

16. John G. Neihardt, *Black Elk Speaks* (Omaha, NE: Bison Books, 1932).

17. Don Trent Jacobs, *Primal Awareness: A True Story of Survival, Awakening and Transformation with the Raramuri Shamans of Mexico* (Rochester, VT: Inner Traditions International, 1998).

18. See the reference list for complete citations of these works.

19. Four Arrows, "The Heart of Everything that Isn't," *Truthout*, accessed 09 June 2018, https://indiancountrymedianetwork.com/culture/arts-entertainment/the-heart-of-everything-that-isnt-the-untold-story-of-anti-indianism-in-drury-and-clavins-book-on-red-cloud/.

20. First People Worldwide, "Who are Indigenous Peoples?" accessed 19 Aug 2018, http://www.firstpeoples.org/who-are-indigenous-peoples.

21. Don Trent Jacobs, *The Bum's Rush: The Selling of Environmental Backlash (Phrases and Fallacies of Rush Limbaugh)* (Boise, ID: Legendary Press, 1974).

22. Four Arrows, Greg Cajete and John Lee, *Critical Neurophilosophy and Indigenous Wisdom* (Netherlands: Sense, 2010).

23. Noam Chomsky, this quotation located on the back cover of Four Arrows, *Teaching Truly: A Curriculum to Indigenize Mainstream Education* (New York, NY: Peter Lang, 2015).

24. David Ferguson, "Arkansas Republican Pushes Bill to Ban Howard Zinn Books," *Rawstory*, 2017, accessed 08 June 2018, https://www.rawstory.com/2017/03/arkansas-republican-pushes-bill-to-ban-howard-zinn-books-from-public-schools/.

25. Bourne, E. G., ed., *The Northmen, Columbus and Cabot, 985–1503: The Voyages of the Northmen, The Voyages of Columbus and of John Cabot* (New York, NY: Charles Scribner's Sons, 1906).

26. Christopher Columbus to King Ferdinand of Spain 04 March 1493. Full letter can be viewed online. Accessed 23 May 2018, http://xroads.virginia.edu/~hyper/hns/garden/columbus.html.

27. Counter-Current News, "Police officers Turn in Badges," accessed online 16 May 2018, http://countercurrentnews.com/2016/11/2-police-officers-turn-badges-support-standing-rock-water-protectors/.

28. Bartolome De Las Casas, *A Short Account of the Destruction of the Indies*, trans. by Nigel Griffen (London, UK: Penguin Group, 1992), accessed online 27 August 2018, http://www.columbia.edu/~daviss/work/files/presentations/casshort/.

29. Ibid., 80.

30. Ibid., 16, 22, 23, 24, 30, 74.

31. Wahinkpe Topa (Four Arrows), *Love that Flows Like a River: The Story of Standing Rock* (New York, NY: Skyhorse [in press]).

32. To read more: Four Arrows, accessed online 23 March 2017, http://www.truth-out.org/author/itemlist/user/48882.

References

Bigelow, Bill, and Bob Peterson. *Rethinking Columbus: The Next 500 Years*. Milwaukee, WI: Rethinking Schools, 1998.

Bourne, E. G., ed. The Northmen, Columbus and Cabot, 985–1503: The Voyages of the Northmen, The Voyages of Columbus and of John Cabot. New York, NY: Charles Scribner's Sons, 1906.

Clifton, James, ed. The Invented Indian: Cultural Fictions & Government Policies. New Brunswick, NJ: Transaction, 1990.

Columbus, Christopher. Letter to King Ferdinand of Spain, describing the results of the first voyage. 04 March 1493. Accessed online 23 May 2018. http://xroads.virginia.edu/~hyper/hns/garden/columbus.html.

Conway, Kellyanne. "La La Land Blunder." *The Guardian*, 3 March 2017. Accessed online 1 August 2018. https://www.theguardian.com/us-news/2017/mar/03/kellyanne-conway-alternative-facts-mistake-oscars.

Cooper, Thomas W. A Time Before Deception: Truth in Communication, Culture and Ethics. Sante Fe, NM: Clear Light, 1998.

Counter-Current News. "Police officers Turn in Badges." Accessed online 16 May 2018. http://countercurrentnews.com/2016/11/2-police-officers-turn-badges-support-standing-rock-water-protectors/.

De Las Casas, Bartolome. *A Short Account of the Destruction of the Indies*. Translated by Nigel Griffen. London, UK: Penguin Group, 1992. Accessed online 27 August 2018. http://www.columbia.edu/~daviss/work/files/presentations/casshort/.

Drury, Bob, Kate Waters, and Tom Clavin. *The Heart of Everything That Is: The Untold Story of Red Cloud, an American Legend*. New York, NY: Margaret K. McElderry Books, 2017.

Edgerton, Robert B. *Sick Societies: Challenging the Myth of Primitive Harmony*. New York, NY: The Free Press, 1992.

Ferguson, David. "Arkansas Republican Pushes Bill to Ban Howard Zinn Books." *Rawstory*. 2017. Accessed 8 June 2018. https://www.rawstory.com/2017/03/arkansas-republican-pushes-bill-to-ban-howard-zinn-books-from-public-schools/.

First Peoples Worldwide. "Who are Indigenous Peoples?" Accessed 19 Aug 2018. http://www.firstpeoples.org/who-are-indigenous-peoples.

Four Arrows [aka Don Trent Jacobs]. The Bum's Rush: The Selling of Environmental Backlash (Phrases and Fallacies of Rush Limbaugh). Boise, ID: Legendary Press, 1974.

_____. "The Heart of Everything that Isn't." *Truthout.* Accessed 9 June 2018, https://indiancountrymedianetwork.com/culture/arts-entertainment/the-heart-of-everything-that-isnt-the-untold-story-of-anti-indianism-in-drury-and-clavins-book-on-red-cloud/.

_____. Love that Flows Like a River: The Story of Standing Rock. New York, NY: Skyhorse (in press).

_____. Point of Departure: Returning to Our More Authentic Worldview for Education and Survival. Charlotte, NC: Information Age, 2016.

_____. Primal Awareness: A True Story of Survival, Awakening and Transformation with the Raramuri Shamans of Mexico. Rochester, VT: Inner Traditions International, 1998.

_____. Teaching Truly: A Curriculum to Indigenize Mainstream Education. New York, NY: Peter Lang, 2015.

_____. Unlearning the Language of Conquest: Scholars Expose Anti-Indianism in America. Austin, TX: University of Texas Press, 2006.

Four Arrows, Greg Cajete, and John Lee. *Critical Neurophilosophy and Indigenous Wisdom.* Netherlands: Sense, 2010.

Four Arrows, and Jessica Jacobs. *Teaching Virtues: Building Character Across the Curriculum.* New York, NY: R&L Education: Scarecrow, 1981.

Frasure, Chris. "Truth in Pre-Han Thought." (2016). Accessed 10 July 2018.

Fronsdal, Gil. "The Perfection of Truth." *Insight Meditation Center.* Accessed 15 July 2018. http://www.insightmeditationcenter.org/books-articles/articles/the-perfection-of-truth/.

Gandhi, Mohandas. "Gandhi on the Meaning of Truth." In *Yeravda Mandir.* 1927. Accessed online 28 April 2018. https://berkleycenter.georgetown.edu/quotes/mohandas-gandhi-on-the-meaning-of-truth.

Joseph, Chief. "On a Visit to Washington, D.C."1879. Accessed online 04 April 2018, http://www.firstpeople.us/FP-Html-Wisdom/ChiefJoseph.html.Keeley, Lawrence H. *War Before Civilization: The Myth of the Peaceful Savage.* New York, NY: Oxford University Press, 2007.

Krech, Shepard. *Ecological Indian: Myth and History.* London, UK: Norton, 2001.

LeBlanc, Steven A., and Katherine E. Register. *Constant Battles: The Myth of the Peaceful, Noble Savage.* New York, NY: St. Martin's Press, 2003.

Loewen, James W. Lies My Teacher Told Me about Christopher Columbus: What Your History Books Got Wrong. New York, NY: The New Press, 2014.

Munro, Donald. *The Concept of Man in Early China.* Stanford, CA: Stanford University Press, 1969.

Neihardt, John G. *Black Elk Speaks.* Omaha, NE: Bison Books, 1932.

Palmer, Parker J. *A Hidden Wholeness: A Journey to an Undivided Life.* San Francisco, CA: Jossey-Bass, 1999.

_____. *The Courage to Teach: Exploring the Inner Landscape of a Teacher's Life.* 10th Anniversary Edition. San Francisco, CA: Jossey-Bass, 2007.

Pinker, Steven. The Better Angels of our Nature: Why Violence has Declined. New York, NY: Penguin Group, 2012.

Thoreau, Henry David. *Indian Notebooks* (1847–1861). Transcribed by Richard F. Fleck. 2007. Accessed online 17 April 2018. https://www.walden.org/wp-content/uploads/2016/03/IndianNotebooks-1.pdf.

———. *A Week on the Concord and Merrimack Rivers.* 1849. Accessed online 5 March 2018. https://archive.org/stream/weekonconcordmer1849thor/weekonconcordmer1849thor_djvu.txt.

Whelan, Robert. *Wild in woods: the myth of the noble eco-savage.* IEA Studies on the Environment 14. London, UK: IEA Environment Unit, 1999.

White, David Gordon. "Yoga, Brief History of an Idea." Accessed 22 June 2018. https://press.princeton.edu/chapters/i9565.pdf.

Zinn, Howard and Anthony Arnove. *A People's History of the United States.* New York, NY: Harper Perennial, 2015.

CHAPTER TEN

Listening to the Trees

TOM MCCALLUM

My given name is Tom McCallum and my spiritual name is White Standing Buffalo. I was born in a small Metis community in northern Saskatchewan. Our community was at the confluence of four rivers and they widened to create a big lake called Isle a La Crosse. Each year the lake froze over and people would enjoy skating near the shore. When I was thirteen years old I was out, along with three other friends, enjoying skating with the other people. I looked at the freshly frozen lake and I suggested that I and the three other friends should go across the lake to an island near the far shore. The people warned us that the ice would not hold our weight and that it was not a good idea. Of course, we were young and did not heed the warning and set off to the far shore.

As we ventured out near the middle of the lake, a crack on the ice had formed and we had to make a choice: turn back, or continue and take a chance that the ice crack would not give when we crossed it. I was the first one to cross it and when it was established that it was safe the others followed. We made it across the lake to the big island and we enjoyed skating there till it started to get dark. I said that maybe we should start for home and when we came to the crack in the ice and the others asked me how we would get across. I said, "The same way we got across before, step over the crack." All the others got across but this seemed too easy for me, so I said that I wanted to jump over the crack. I skated back and took a run at the crack, and jumped over.

When I landed on the other side, I broke through the ice and immediately sank under the water. I looked around me and it was pitch black. I looked up and I could see light where I had broken through. I started to swim for the light and finally surfaced. The cold air hit me and I was gasping for air and tried to climb out of the water. I could not get on top of the ice as the ice would bend and I would slide back into the water. One of the small guys jumped into the water saying, "I will help you," but then he was also in trouble. The other small guy lifted him out of the water and they both headed for shore. There was one guy left, the biggest guy in our group. He tried to come toward me to help but each time he got close the ice would crack. Eventually he said, "I will get help" and he skated toward town. I was left alone struggling to stay afloat and on top of the ice. I was exhausted as I kept sinking and swallowing water, and the last time I sank I knew I was drowning and had the sensation that everything was ok. I had no sensation of being cold or of being under water. It felt like I was somewhere else. At this time I had a vision of a trembling aspen or poplar tree, with its leaves fluttering in the wind. I became very sad and felt like crying because I would never have the opportunity to see a poplar tree again. I don't know how long this lasted but it was as if I came awake and thought that I could not let this happen.

I started to swim toward the light once again and resurfaced. As I got on top of the water I reached out and found a perfectly round hole in the ice which was not there before. I hung on to this hole in the ice and called for my friend who was a long way away. He seemed to not hear me so I screamed at the top of my lungs and he stopped and looked in my direction. I called for him to come back and he hesitated and I screamed at him again to return. He started skating back toward me and when he got close I told him to throw me his coat. He tried throwing his coat but he was too far away for me to reach it. I told him to come closer and he did, this was when I was able to reach his coat and he pulled me out. I let out a loud whoop as I was being pulled out of the water and to the top of the ice. This was the beginning of a long journey with the trembling aspen, which is still continuing to unfold.

When I was about 39 years of age, I made offerings to a medicine woman and told her the story of drowning and asked her why I saw the trembling aspen instead of my mother or some other relative. She prayed with the offering and then said to me that this tree was the Sundance tree, that it had saved me, and that someday I would have to repay this tree for giving me another chance at life. I attended the Sundance ceremony that year and helped out.[1] The following year I was called by the tree and I started dancing. This was in 1986, and although I had danced, I really had not formally given thanks to the tree for saving my life. In 1994, I brought a long blue cloth, tobacco, a blanket, and a monetary gift to the Sundance and spoke to an Elder. I relayed my story and he said, "Come with me."

I stood with this Elder, along with five other Elders, as the people raised the Sundance tree. As it was being raised all the Elders started to cry very loudly. I also started crying, though at the time I did not know why. When it was fully upright everyone gave out a loud cheer. On the last day of the ceremony, during the dancing I was transported to somewhere else and had no sensation of dancing. I could still hear the drums, the whistles, and the people talking, but the place I experienced was not the Sundance. I saw a beautiful place where other people were walking and talking in very gentle tones. It was so serene and the grass was a brilliant green as if it was made of crystal. I spent all day at this place and did not want to return, I had found what most people would refer to as heaven.

I continued dancing every year except for a few years at which time I was unable to attend. Around 2004, I was invited by a Sundance leader to go and dance at his Sundance. I agreed to dance for one year, however, when I got to the Sundance I was instructed that I had to dance for four years. I was not very happy about that, but I was told that it was for a very specific purpose. When I finished this commitment I said that I was not dancing anymore, I had done enough. I attended a ceremony and the spirits told me that I was to host a Sundance. I had dreamed about an Elder telling me to take a branch from a Sundance tree and plant it somewhere else. This dream had come to fruition; I was now a Sundance chief.

I have since come to understand that the ceremony is one of renewal and sacrifice so that other people may live. The tree of life or center pole gives up its life, so that we may get healing and reconnect with our spirit along with shedding our burdens which we may have picked up during the year. The sacred energy that is generated by this tree of life is incorporated into each attendee at the ceremony. If you can recall an experiment that is done at school with a magnet, a piece of paper and iron filings, the pattern that the filings display is the same pattern of energy generated by the sacred tree. People are encouraged to go to the tree to ask directly for anything that may assist them on their spiritual journey. At this time people may ask for names or special healing to occur for themselves or their loved ones. As I said earlier the trembling aspen has a certain biochemistry and biology which facilitates the transfer of this sacred energy from the star nations, which I will return to below. The tree does not work in isolation, but is a vital link in the transfer of sacred energy to the participants and attendees.

The tree is decorated with multi-colored cloth to represent our prayers and all that we have come to ask. The colors of this cloth represent the vibrational frequencies of the universe. Our part in this is that we need to offer cloth and tobacco to access the portal or energy field through which the flow of spirit will enter our being and facilitate healing. This ceremony is available to anyone who wishes to transcend the limited perception of life and engage in bringing forth the true essence of reality. Just as we wear our finest in the presence of the Creator, so too is the tree of life wearing its finest for its final journey to the Creator. After the ceremony is over the tree of life will remain standing four days after which time it will rejoin the Creator. The Sundance chief will remain with the tree at this time. The term Sundance is a contemporary word that has been applied to what the Cree refer to as "*pahkwisimowin,*" which translates to thirst dance. The dancers give up all food and water for the duration of the ceremony in order to sacrifice themselves and purge themselves of any toxins which may get in the way of their connection to the sacred energy. In this way, the dancers concentrate on the spiritual aspect of themselves and are not distracted by physical restraints.

The Sundance ceremony is very old, although there are stories about when it was given to the people; no specific date has been established as to its transference. The reference is to a very ancient time. This is one aspect of renewing the connection we have to all the other aspects of the Hierarchy of Dependence, a spiral model that depicts how we are connected to all of the universe. We were given ceremonies, to reconnect us and renew our link in all of creation. This special ceremony gathered people yearly at specific times to generate and rekindle a vortex which, when activated, allowed the participants to access the energy from the universe (or field as it is sometimes called), to replenish or recharge humanity. This ceremony is comprised of all the various aspects of the model of dependence, the people, animals, plants, spirits and, of course, Creator. Through the re-enactment of the Sundance, the people orchestrate the concept of universal sustainability by including all of creation. In addition to the people being recharged and renewed, the sacred energy or spirit also is mobilized and a very large energy field is activated which will reverberate through seven generations in all levels of creation. All ages of humanity are involved in this event. The children have an intricate role to play in this ceremony. Children are recognized as pure spirit as they have not been contaminated by living in society. The Cree word for child is "*awasis,*" when this is translated in its most profound sense it means, "being of light."

The sacred tree which is planted in the center of the Sundance arbor represents among other things the center of the energy field. It is much like the core of an electromagnetic coil which generates an electromagnetic field when electricity is connected. The energy comes from the people and also from all the elements of the Sundance lodge which of course includes the spirits or energy field. The songs that are sung and the dancing which takes place raise the vibrational frequency of the lodge and in all of the people. This energy radiates outward and travels through time and space for seven generations into the future.

The day we call tree day starts with a gathering of branches of various willows. This is called the gathering of the thunderbird nest; it is done as the dawn is approaching. A leader has been selected who will gather seven men who are orphans and they will go into the bush to gather these branches. Upon their return they will be wolves who howl as they approach the Sundance area. A pipe ceremony is done with these seven men at the front of the proposed lodge.

The men will then gather to select a tree which will become the center pole. Prior to the pipe, permission is asked of the tree's father and mother to offer its son for the Sundance. A pipe ceremony is held at the tree to honor the tree for giving up its life. The Sundance chief sings to the tree to give life back to the tree via the pipe and song. After this ceremony is over, the Sundance chief leaves the area and the men will start to chop the tree with special axes that have never touched the ground.

When the tree is down the men will carry the tree to the arbor where blankets have been laid on the ground to lay the tree on so it will not touch the ground. The tree is then carved with lightning bolts and a buffalo and thunderbird are carved on the east and west side of the tree. The tree is then decorated with all the cloths that are 13 meters in length. The nest is then put on the tree and the tree is ready for raising. After the tree is raised and secured, one meter length of white cloth is filled with tobacco which is then offered in the form of an egg which will be put into the nest.

When this is finished the rest of the lodge is ready to be built. This may take up to four hours to complete. In a radius of 22 feet, Y shaped poles about eight feet in length are put into the ground which will act as supports for the rafters that intersect at the center pole. These Y shaped poles will number 12 and will be equidistant from each other. After the rafters are put into place, stalls will be made for the dancers.

The dancers will have their last meal and then will have a sweat lodge ceremony. They will enter the Sundance lodge and will begin dancing when everyone is inside. The dancers will dance till about midnight then they will get some sleep. Before sunrise the next morning dancing will resume and various ceremonies will take place during the day within the arbor. At about 10:00 or 11:00 p.m., the dancers will once again retire and have some rest.

The Elders have said that this is the closest we will come to being in the presence of the Creator.

Guidance from the Trembling Aspen

The future of generations to come is at risk. We can use some guidance for making choices to help prevent them from being plagued by the tragic results of our current imbalanced relationships in the world. As a messenger who has been tasked to provide such guidance to the general public, I offer this presentation. I describe the challenging journey that helped me understand my dreams and brought me to this role. I tell how other-than-human teachers who helped clear the lens through which I see existence have asked me to help others similarly. I explain why non-human assistance is vital, and how ceremonies I conduct use the unique qualities and chemistry of the Trembling Aspen (White Poplar) to reveal sacred metaphors and symbolic meanings that can help us. The ideas imparted to me by this sacred tree facilitate the transference of information from the Star People who claim they are our older brothers and sisters who wish to assist us on our journey at this crucial time in history.

Messages from the Star Nations

In the Cree language we do not have a word to describe spirituality. The word we use to describe this connection is "*Kaki isi miko isit nehiyow*," this translates to, "The gifts we were given as a four directions people." We also have a word "*pima-tisiwin*," that translates to life. But at a much deeper level the word is made up of three words. "*Pi*" means to come or to unfold. "*Mati*" comes from the word "*maci*," which means to start or the beginning. "*Siwin*" is the action. So in essence this word means life or the unfolding, as the physicist David Bohm said.

What is the universe trying to tell us? The world is over-populated and there is a lack of resources to sustain the exponential growth. The animals are moving into the cities and towns because we have decimated the natural habitat. Our thirst for more of the same has imbalanced the ecosystem and we have stretched it beyond its limits. Still, we continue, in spite of this knowledge, we deceive ourselves into believing that science will give us a magic wand that will reverse this depletion.

Perhaps we will find a new earth somewhere else where we may continue to try to quench our desire to fulfill the void we feel inside ourselves. The Lakota people have a word to describe this wanton desire, "*wasicu*," and it translates to "eaters of fat." How do we address this trance-like state that captures and imprisons our minds? How do we replace the lens through which we see we see life?

Is it even possible to counter the damage we continue to wreak upon the ecosystem and reverse this destructive trend? We seem to be caught in this whirlwind of belief that all will be well in spite of our headlong rush toward self-destruction. Is there something humanly possible that we have not considered? What will it take to shock us out of our current perceived reality and assist us to change our perception of sustainability and ultimate survival?

In a book called *The Hidden Life of Trees*, German forester and author Peter Wohlleben (born in 1964, in Bonn, Germany), who writes on ecological themes in popular language, asserts that the trees have feelings and also communicate to each other and others who listen. They have family systems similar to human beings and they support each other in sustaining life.[2]

In the Aboriginal world, we have a system whose foundation is balance. This concept of balance starts within everyone. The heart has four chambers that symbolically represent the four aspects of life, Spiritual, Emotional, Physical and Mental. This concept extends outward to include our family, community, province or state, country, and so on.

In the Cree culture, we have what we call the *hierarchy of dependence*. This relationship is represented by a spiral (shown in Figure 10.1). The spiral represents that we are not disconnected, but that we are intricately connected to all of creation. In the center are the people, next are the animals, then the plants, then the minerals, followed by the spirits, and eventually the Creator. We are in the center—not because we are the most important, but because we are the most dependent. We depend on all the rest for our survival. All the rest do not need us for their survival, and if all the people were gone, everything else would thrive.

Notice the similarities among the Hierarchy of Dependence, the Fibonacci spiral, and the designs of fractal geometry in nature. In the world of fractals, Konstantin Eriksen, wrote a piece titled, "The Astonishing Geometry of Nature," for the online blog Wake Up World, "To me, the obvious conclusion is that there is an inherent form of intelligence in nature. In a sense, the cosmos has a form of consciousness. On every scale imaginable, from biological cells, to ferns, to mountain ranges, to galaxies, nature finds a way to self-organize amidst the apparent chaos."[3]

The Fibonacci spiral is found in most of nature, and if one looks at the Hierarchy of Dependence, it is also in the pattern of this spiral. It would be very interesting if one were to research the Hierarchy of Dependence and how it relates to the Fibonacci spiral and to fractal geometry. Benoit B. Mandelbrot in his book, *The Fractal Geometry of Nature*, claims that nature has a complexity in the way that shapes differ in kind, not merely degree.[4]

In terms of sustainability the spiral shows that we are connected to all of creation and that what we do to one aspect in this spiral will naturally affect the other links seven generations into the future. The people who were created were given a mind and free choice to make abundant the veritable Garden of Eden upon which we were born. All of the rest of the inhabitants in the spiral or creation encircle and protect us and are ever ready to assist us whenever we call for their help. The reason we are in the center is because all of creation starts from the Creator who is omnipresent. We were made last and the Creator loved us so much that he did

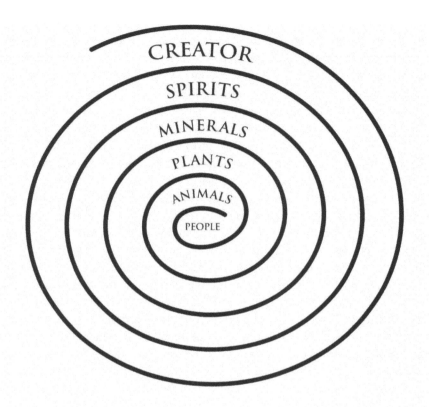

HIERARCHY OF DEPENDENCE

Figure 10.1: Hierarchy of Dependence.
Source: Courtesy of Saskatchewan Indian Federated College.

not want to let us go until he was certain that we had helpers who would assist and protect us. All of the other aspects of the hierarchy of dependence within the spiral were put there by the Creator to sustain the human race. These are all the rest in the spiral. As long as we respect all of the others and listen to their instructions we would continue to thrive and make a stronger connection with our relatives and elevate our sacred energy or spirit and eventually transcend the material world.

We are a vital part of this link in all of creation and we were the last to be manifested. As a result of this, we are like two-year-olds learning to navigate through life and not really paying heed to anything else except our own exploration. As the youngest beings in life, we have a very limited understanding of how we are intricately connected to all of creation. It seems that in all our encounters we have this strange desire to control, with the result being a disruption of the natural flow of life or energy. This disruption creates an imbalance and our lives become toxic; leaking a poison to all the other links in the web of life.

The second level of the spiral holds the animal kingdom. The animals agreed with the Creator to help out their younger siblings, the people. Each animal had a specific duty that he promised he would share with the people, as long as the people would continue to follow the protocols that were set in place to retain balance and continue the growth of life. Our closest relative, the bear, agreed to be the healer and provide the people with medicine from his body. When we honour the bear in our ceremonies by feeding its spirit, it comes and presents itself and offers healing to whoever is in need. We have a special ceremony that is a part of the Sundance where the bear dancers come to heal the people. One of the spirit medicines that we garner from the bear is its fat or grease, which is used for many preparations. As long as we continue to respect this gift of life that is offered to us by the bear, it will always be there whenever we call for it. The bear represents the gift of the western doorway in the Medicine Wheel, which is the unknown or great mystery of life. This unknown represents the journey we take from our minds to our heart. It is a very difficult journey that takes a lifetime to accomplish. We live in a very mental world and find it difficult to go to the heart; becoming a caring person is seen by our society as being soft or weak. Only by connecting to the heart as part of our journey to wholeness also known as the emotional aspect can we sustain ourselves and all of creation. This grandfather, as we call the bear, is one example of our connection to the animal world or kingdom. Each animal also has a great gift that it offers us as people and when we follow the instructions that they provide, they, like the bear, will continue to offer their particular gifts, and others, to us.

The next level on the spiral is the plant relatives; both the people and the animals are dependent on this species for their survival. Each variety of plant life has the ability to restore balance in its own way. At the time of creation, the plants offered themselves to the animals and people to be used as needed. The animals still carry the original instructions; however, the people have forgotten most of the relationships or connection to this aspect of creation.

An example of the plant kingdom is the willow. The willow is the strongest plant known in the forest. It can withstand the severest cold in winter. Other trees will crack from the cold, but the willow will continue to grow. A fire will destroy other trees, but the willow can withstand the fire and continue to thrive. You can even cut the willow and it will not stop growing. The flexibility of the willow teaches us about resiliency, the ability to bounce back from all adversity and continue on our journey. The most sought after gift from the red willow is the ability to express yourself openly and honestly without harming anyone or yourself. The bark of the red willow has a property called acetylsalicylic that is used in aspirin to address headaches and is also used as a blood thinner. These properties of the red willow are gifts that have been given to the people and as long as we respect the plant world and do our part, they will always be there for us. We are in a unique position to care for the sacred gifts of the plant kingdom, in doing so we will truly be practicing sustainability.

The next level on the spiral is the minerals. Our mother, the Earth, is referred to as a mineral. The earth is like a huge garden from which we reap all that we need to sustain us in our growth. Like a young child being held to the breast of the mother and nurtured, so is our relationship to our mother, the earth. The earth and all that grows on it is finite, as long as we take care of our mother, she will continue to nurture us and the generations to follow. Incidentally, the grass that grows on mother earth is referred to as "*muskosi,*" in the Cree language. This translates to "hair of the bear." In other words, the earth is a bear, and since the earth is represented as a bear, and the bear represents healing, then the earth is seen as a healer.

In the Alaskan community of Haines, a man started to experiment with immersing his feet into the earth. After a period of time he noticed that the ailments he had previously felt were absent. He was not sure if the earth really affected him or if it was something else, so he asked his friend to try putting his bare feet on the ground. Amazingly, his friend also experienced a relief from ailments. The man had other people from the community try this experiment and to his astonishment, the energy of the earth provided relief to all of the people. One man who had been in a wheelchair for 15 years participated in the experiment, but his way was a little different from the rest. He had a probe inserted into the ground and a wire run to a pad inside the house. When this man put his feet on the pad, he felt amazing results and was eventually able to move his legs. In time, he was able to walk with the assistance of supports.

Word got out and eventually the man who had begun the experiment was called to attend talk shows and share his experience in addition to the community impact. When this news got out, the astronaut Dr. Edgar Mitchell along with Dr. David Suzuki got involved with this grounding experience. They now have a theory called "grounded theory," to explain the results of this exercise.

Another account of this energy and how it interacts with the earth comes from a DVD called "Thrive," which is the culmination of lifelong research conducted by a man called Foster Gamble. In the DVD he describes how the energy field works on the earth. It comes from the North Pole, goes to the equator, and returns to the North Pole. The same happens for the South Pole.

The next level is the spirits (or energy, as it is called by the scientists). Our bodies are infused with energy and we are a part of the whole energy field or universe. This tells us that we are intricately connected to this level. In the Aboriginal world we have ceremonies which assist us in renewal and a clearing of blockages caused by our stressful encounters with each other and interactions with negative thoughts. One such ceremony is the Sundance, where people gather together with one heart and one mind and sacrifice themselves so that others may live and be healthy. The earth, trees, sun, and wind are all part of this ceremony and all contribute to assist us in raising our vibrational frequency. When we do this, it opens a vortex so we can experience healing and shed ourselves of the negativity we have picked up and carried during the year. The ceremonies we have been given are the vehicles by which we make connection to the spirits or energy field. Vibrational frequencies are changed, and as a result, we have access to the other levels or fields of potential. These ceremonies have very specific orders or methods that are used to make this happen. Not all people have been gifted with ability. We have very special people we sometimes refer to as Medicine people or healers who can activate this process. This is one reason we respect our Elders; they are our conduit to the sacred energy. As people, our part in this level is to acknowledge and follow the direction and instructions we have been given by this sacred energy of spirit. As long as we do our part and leave ego aside, the spirits will be there to assist us whenever we call them.

We need to break free from our myopic worldview and entertain the possibility that maybe there really is another perspective that may contribute to a restoration of balance. There is good news in all of this doomsday rhetoric. There is help that we can access which will assist us in returning to a more balanced, harmonious, and life sustaining system. Aboriginal peoples have been given certain ceremonies to contribute to the restoration of balance. These ceremonies are like the scientific experiments conducted by learned individuals. Aboriginal ceremonies are not conducted in isolation; they call forth the energy or spirits of all to be a part of the event. We do not eliminate variables; rather we welcome all of the sacred energies or spirits to take part.

Returning to the Original Instructions

I (White Standing Buffalo) have been tasked to present an alternative solution, which has always been a part of creation and is available to everyone, as a viable opportunity. The sacred energy or field, as it is called by physicist Lynne McTaggart, holds all the knowledge ever since the dawn of creation.[5] Within this sacred energy there are formulas written as symbolic patterns superimposed on the DNA of certain trees, to which we have access. Within this energy field lays the answer to our current dilemma. All we need to do is listen and heed the instructions we will be given. In the words of the sacred energy, as given to me in 2016:

> We must return to the original instructions to acquire the assistance of other than human beings. These instructions have been carried on by a few people who have not strayed or have returned to the interconnected world-view. Time is of the essence, we must incorporate all the knowledge and work together to restore this life to one of self-sustaining instead of a deliberate road to self-destruction.[6]

So, how do we do this? The messages from our other relatives must not fall on deaf ears. The other star peoples are constantly relaying messages to certain trees, who in turn relay these messages to us. Do not wait till it is too late to clean up the environment. Let's work together to restore balance and allow the other aspects of our gift, the earth, to replenish and thrive for ourselves and future generations. The quantum world-view suggests that we are all connected and have conducted experiments to support this view, in the words of our elders, *What we do to one part of the web of life we do to ourselves.*[7]

These two views support each other and I believe that they are on the right track to a flourishing world, where there will be benefits to all the people and a return of prosperity. Following are the instructions of the tree people as imparted to me:

> The standing ones have many, many geometric designs within their being that are tied to outer galactic beings. They are signatures that allow for the communication to occur within. They are maps that have been imprinted into the physiology and biochemistry of the standing ones, along with all of the other aspects of nature. As you already know nature is on that list, however, there are very specific repetitive diagramming that occurs with an occasional imprint of another diagram, for human beings this is linked to the DNA. The sequencing is very similar, carbon, nitrogen, hydrogen, and oxygen, the sequencing that all of nature is bound by this particular alchemy. When it relates to the standing ones, they are in essence holding the kin and connection to the outer galactic beings. This is why you have been instructed very specifically to utilize the star lodge. It is through this mechanism, communication occurs and there is transference of knowledge that is necessary for the time being for all of humanity.
>
> When you take the opportunity and set aside and devote time, very specifically to sit and be amongst these beautiful beings, the standing ones, it provides them the opportunity to work with your own biogenetic systems and your DNA in resetting but also correcting any malfunctions.
>
> There was much that occurred to your physical being when you were having your time with them. They reset many elements of your physiology, your biology, and your

DNA, that had over time become compromised as a result of your life experience. This will continue to unfold over the next four to six months. There will be many, many healings that will occur internally which you will not even be aware of, however, there will be periods of time where you will experience high energy and very low energy. And there will be times also when you will feel like you absolutely must sleep and will sleep for hours. It is during this time when there is uploading of new information into your physiology or your biology. You must honour this as this is where the geometric designing is being imprinted to your own being. Very specifically, you are given this direction to take notes when these occasions are occurring; as this is what you will be sharing in your presentation. Your own physical effects of this experience with the standing ones and the star lodge ceremony specifically. You will actually be relating to the audience, information that has not been experienced totally by the psyche of the mental being of humanity.[8]

The prophecies from the Aboriginal world tell of what lies in the future. These prophecies are not conducive to a vibrant environment, but we have time and the ability to change what is possible. An example of a prophecy which has unfolded follows.

In the early 1800s a meteorite fell from the sky, the elders had a ceremony with this meteorite and it told them that they were in the midnight of their lives but that they would come out of the midnight of their lives when the eagle lands on the moon. In 1969 when astronaut Neil Armstrong landed on the moon his first words to earth were, "The eagle has landed."

The Aboriginal people became awake politically in 1969 and formed organizations to address the lack of services they were experiencing. One of the prophecies was that the Aboriginal people would be leaders of tomorrow. That tomorrow is here and we need to all work together to address the imbalance in our environment. Our ceremonies will give us the guidance we need to rectify this impasse we are experiencing.

What is your part in all of this? Do you think we should give this aspect or beings some of our attention and see through a different lens for the possibility of sustainable living where everyone can enjoy the gifts of the Creator? According to Gregory Cajete, sustainability is honoring the privilege of being alive.[9] In the Cree culture, knowing our connection to all of creation starts with the traditional name that we are given, understanding this name acts as a guide on our journey toward wholeness or following the original instructions. Our ceremonies are connections to the source and they revitalize us and infuse us with sacred energy in addition to receiving further instructions to fulfill our commitment to Creator. As long as we maintain this connection in the spiral of life and we do our part we will have a sustainable creation.

The choice is yours.

Discussion Questions

How does this work comment on the role of senses, both literally and figuratively?

How do the elements of ritual and community contribute to the Sundance ceremony? How might they contribute to the concept of healing on a larger or smaller scale?

Is sacrifice a burden or a lifting of burdens (or both) for the entity making the sacrifice? For the beneficiary? Explain.

Explain the Hierarchy of Dependence model. How does this model communicate sustainability? How does it communicate wisdom? Explain.

Notes

1. The Sundance Ceremony is a ceremony of healing, where people gather together, making personal sacrifices, so that others may live and be healthy.
2. Peter Wohlleben, *The Hidden Life of Trees*, trans. by Jane Billinghurst (Vancouver, BC: Greystone Books, 2016), originally published as *Das geheime Leben der Bäume: was sie fühlen, wie sie kommunizieren—die Entdeckung einer verborgenen Welt* (Munich: Ludwig Verlag, 2015).
3. Konstantin Eriksen, "The Astonishing Geometry of Nature," Wake Up World [blog], accessed online 31 August 2018, https://wakeup-world.com/2012/04/04/the-astonishing-geometry-of-nature/.
4. Benoit Mandelbrot, *The Fractal Geometry of Nature* (New York, NY: W. H. Freeman, 2006).
5. Lynne McTaggart, *The Field: The Quest for the Secret Force of the Universe* (New York, NY: Harper Collins, 2002).
6. Instructions given to me by the Spirits in a ceremony in 2016.
7. Attributed to Chief Seattle, my emphasis.
8. This symbology is linked to my dreamtime, and again, details knowledge and instructions imparted to me by the Spirits through ceremony.
9. Gregory Cajete (presentation at the Sustainable Wisdom Conference, Notre Dame, Indiana, September 2016).

References

Cajete, Gregory. Presentation at the Sustainable Wisdom Conference. Notre Dame, Indiana, 2016.

Eriksen, Konstantin. "The Astonishing Geometry of Nature." Wake Up World [blog]. Accessed online 31 August 2018. https://wakeup-world.com/2012/04/04/the-astonishing-geometry-of-nature/.

Mandelbrot, Benoit. *The Fractal Geometry of Nature*. New York, NY: W. H. Freeman, 2006.

McTaggart, Lynne. *The Field: The Quest for the Secret Force of the Universe*. New York, NY: Harper Collins, 2002.

Wohlleben, Peter. *The Hidden Life of Trees*. Translated by Jane Billinghurst. Vancouver, BC: Greystone Books, 2016. Originally published as *Das geheime Leben der Bäume: was sie fühlen, wie sie kommunizieren—die Entdeckung einer verborgenen Welt*. Munich: Ludwig Verlag, 2015.

Steps Toward Integrated Futures

CHAPTER ELEVEN

Connection Modeling

Metrics for Deep Nature-Connection, Mentoring, and Culture Repair

JON YOUNG

Acknowledgments

Gratitude

I give thanks for:

- All of our ancestors who have been supportive, nurturing, and connective; keeping their shoulders back and modeling how to become fully vital.
- All those helping with healing and supporting life.
- All those who have helped me recognize how to supercharge our Connections and capacities.
- The future generations of people and nature.
- My awareness of every child on the planet now yearning for Connection with the adults who support them.
- My own personal connections, the fullness of my days, and my own health and wellbeing.
- The creation of core routines, structures, and containers that bring out everyone's best; including health, happiness, unconditional listening, empathy, helpfulness, sacred aliveness, love, quiet mind, gift and genius.
- The comprehensive healing of our social systems for the health and wellbeing of all people, especially all the children in the world.

Appreciation

- To the Pokagon band of the Potawatomi people, the local folk who have held the land near the University of Notre Dame since the beginning of time, and to all the people who stand behind them for all time.
- To Darcia and Gene, and everybody involved in holding the Sustainable Wisdom conference. Thank you for including me.
- To everyone who presented at the conference before me who was so inspiring.
- To everyone who I haven't mentioned because they definitely deserve to be mentioned.

Lineage

- My parents, New Jersey, Ruth and Arthur
- Nanny Cecil, Irish trickster-storyteller, paternal grandmother
- Maternal Great Aunt Carrie, Poland
- Tom Brown, Jr., tracker, coyote mentor, and outdoor survivalist, Pine Barrens, New Jersey
- Stalking Wolf, Apache, my mentor's mentor
- M. Norman "Ingwe" Powell, elder of British ancestry, initiated and raised as Akamba and Zulu, co-founder of Connection modeling, who joined me in 1983 and worked with me and many others until his passing in 2005
- Jake & Judy Swamp, Kanien'kehá:ka (Mohawk) elders, Akwesasne Tree of Peace Society for their guidance in peace-making ways
- Mike McDonald, Kanien'kehá:ka (Mohawk), Tree of Peace Society for helping me to understand history from a different lens
- Gilbert Walking Bull, Pine Ridge Lakota healer, elder to me and consultant to our connection work until his passing in 2007
- Paul Raphael, Anishinaabe and Odawa peacemaker, Peshawbestown Michigan for his contribution to our understanding of the role of elders and grieving to heal dis-connection
- Ganuma, Naro San Bushman elder and healer, one of the most profoundly connected people I have known

Overview

Human beings are facing a great connection challenge. The prevailing worldview, based on many generations of ancestral trauma, has resulted in, and perpetuates, the current epidemic of separation. This separation has resulted in widespread repercussions to the body, mind, spirit, and the natural world. To counteract separation, I propose a process of re-connection to oneself, to other people, and to nature. The restoration of this connection is critical to the continuance of life on Earth.

My research has focused on universalizing and applying effective models for re-connection. We now know that a deep and meaningful relationship with nature is a critical part of restoring connection. Yet Nature-Connection has remained elusive for many modern institutions, communities, parents, families, and individuals. Most current approaches have failed to increase connection in a meaningful way. Educational, recreational, and wellness-focused models are well researched, supported, and established, but connection models have been largely misunderstood.

Simply going outdoors, though helpful, is not enough to become Nature-Connected, which requires its own tools and methodology to be truly beneficial. Just as we cannot exercise our biceps by reading about them, or learn advanced mathematics by jogging, we cannot use purely cognitive or recreational approaches to increase connection effectively.

For over thirty years, I have designed collected techniques, designed and tested models, trained others to use these models, and have led workshops facilitating connection models that have proven dependable and predictable in their effectiveness. Drawn from earth-based approaches, these connection models are ergonomic in their application—they are neither philosophical nor are they based on any particular belief structure or lineage. These models correspond to recent neuro-biological evidence that foundational learning and development in nature is our universal human "blueprint." Based largely on the intergenerational transfer of skills and connection-based knowledge, I call this approach "Cultural Mentoring and Connection Modeling." This model results in "cultural repair"—since one of the primary functions of a culture is to provide for connection—and it aids in the reestablishment of a thriving, regenerative culture for future generations.

When connection modeling is understood and applied, and when the outcomes are measured in terms of connection, it results in what I call "Deep Nature-Connection"—a relational interconnectedness that transcends the separation of the self and nature. Deep Nature-Connection reliably results in eight measurable attributes that arise in the connected individual. These attributes are listed below and are graphically represented in Figure 11.1, below.

The Eight Attributes of Connection

1. Quiet Mind: presence, unbridled creativity based on sensory integration, access to one's unique genius.

2. Inner Happiness: as of a child, innocent and fresh, with glee, wonder, and brightness.
3. Vitality and Abundance of Energy: one elder described this as an abundance of electricity in the body.
4. Unconditional Listening and Mentoring: capacity to catch the stories of others, listening deeply without judgement or need to advise, commitment to mentoring and "paying it forward."
5. Empathy: feeling into our connection with nature, deeply feeling this connection in our own bodies leads to love and respect for nature .
6. Truly Helpful: personal gifts and vision are surfaced and activated, initiative, service to others.
7. Fully Alive: awareness of the sacredness of life, a sense of awe, respect, and wonder.
8. Unconditional Love and Forgiveness: compassion and forgiveness, and understanding with respect to the people of our world, forgiveness as a core routine.

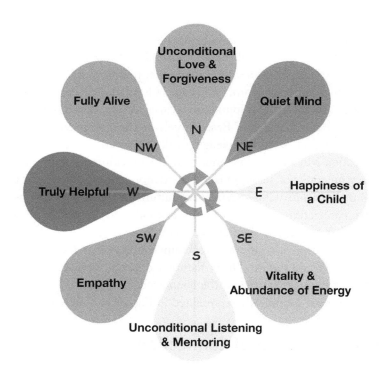

Figure 11.1: Eight Attributes of Connection.
Source: Author.

Origins

When I grew up, I was the last child in the woods, just like in Richard Louv's book about saving our children from nature deficit disorder.[1] Television only had three channels, so I went into the woods and played a lot with other children of different ages, avoiding adults, and taking risks without supervision. A big part of my life growing up was during the timeless and self-driven time before media got too strong and before the culture shifted so that children no longer connected to nature. The research now shows that this was incredibly important to my evolving neurology.[2] And nowadays very few children play like that.

A few remnants of healthy culture from Europe persisted in my family lineage. My grandmothers had immigrated to the US from Ireland and Poland, and they both had tricks up their sleeves. Nanny Cecil (from Irish lineage) would send me on errands to go find things in nature and bring them back to her, which could result in me exploring for weeks. My late friend and mentor, Gilbert Walking Bull (himself a traditional wisdom keeper and healer), told me his grandmother did the same thing with him, sending him on errands. This is the kind of cultural element that we need to restore.

My adopted uncle, Gilbert Walking Bull, passed away and became an ancestor in 2007. He called me his nephew. We made a mutual commitment that I carry on now. He would help me with my work, if I would help the people in his community, especially the youth in Pine Ridge, where the teen suicide rate is incredibly troubling. I have worked with Kathleen Lockyer, an occupational therapist, for the last 15 years to better understand why my techniques work and how the neurological system grabs a hold of them so readily. Many kids—about 1 in 29 children—across the United States, are suffering from sensory processing disorder.[3] In Silicon Valley it's more like one in ten. And the teen suicide rate is linked directly to this. Our children are in deep trouble. So, it is my commitment to do good work for these children, on behalf of this elder and others who have done so much for me and for my family.

In New Jersey in 1970, I felt like one of the only children who was witnessing the tide of life receding. That's one of the formative things that still motivates me. When I was 10 or 11 years old, I noticed snakes disappearing from the habitat, or turtles not laying eggs anymore. When I tried to get other people concerned, I had a lot of difficulty. People who love nature will care for nature. And we need more human beings to care for nature, don't we? But only people who connect to nature will love nature. If people are not connected to nature, they won't take care of it. It doesn't matter how many laws, or regulations, or agreements we try to impose externally. We need more humans who intrinsically to want to take care of nature. And the only way I've found to do that is to connect them to nature.

So, I began a line of inquiry as an anthropology student in 1979. What does Nature-Connection mean? How can we foster the results of Nature-Connection predictably and scaleably? Who in the world is the most connected to nature, and how did they get that way? Why are some people connected to nature and others simply are not?

My research revealed that the majority of people who were not connected to nature were from Westernized society, or, if not from a Westernized society, their society had been affected by historical trauma. Historical trauma has become integrated with cultural behaviors, passed on through time, and normalized to become mainstream behaviors. What we call modernism is saturated with normalized, socially-accepted, historical trauma.

The people who are connected to nature are those who are still Indigenous. Among Indigenous people around the world, some are incredibly effective at getting a high percentage of their children connected to nature at a very deep level. Others only have a select number who are connected so deeply. So I asked, why do some groups only look to initiate and deeply connect a few people?

Gilbert Walking Bull, who grew up homeschooled in Indigenous culture by his parents (both traditional healers), explained this to me, saying,

> In Lakota culture, we look for a certain number of people in every generation. The ones who we're going to initiate into holding and understanding how to regenerate the culture. We raise these children very close to nature. And when they get to a certain age, we start initiating them into ceremony and ritual. So they begin to hold the culture as they get to be elders.[4]

Other cultures had a mandate to ensure that nearly every child was Deeply Nature-Connected. I noticed a high correlation between difficult places to live and this kind of cultural mandate. The San Bushmen inhabit a really difficult place to live, with unbelievably numerous ways someone can die. So if they don't raise their children deeply connected to nature, they would be dooming their children. Their educational model relies on first connecting children deeply to nature, and then letting the children take care of themselves. The San Bushman now have a reputation for being the best trackers in Africa, and nearly all their children are Deeply Nature-Connected.

All humans come from the San Bushmen. Recent genetic mapping says the San Bushmen are our ancestors.[5] They're still alive, with a healthy culture like we all had for two million years, before the onset of modern society. The San Bushmen retain a lot of cultural wisdom supporting how all human neurology is designed. We are all hardwired biologically, even modern humans, to be aware and Nature-Connected.

I began another line of inquiry: what were the traditional San Bushmen doing that the traditional Apaches, Lakotas, Aboriginals in Australia, and other native hunter gatherers were also doing? What do the ancient stories of Ireland say we used to do? What are modern people not doing anymore that they could be doing and would be helpful to support connection? Maybe there's a clue there. What elements are shared by Indigenous hunter-gatherer cultures that produce Deeply Nature-Connected trackers? I excluded elements related to warfare, using only elements with peaceful purposes. In my research I started noticing, compiling, and correlating a growing number of these shared cultural elements, looking for an ergonomic application. For instance, greeting customs are consistently important among Indigenous people who raise deeply connected people. San Bushmen greeting customs can last up to 45 minutes, and often occur several times a day.

In 1983, when I experimented with integrating just one of these cultural elements, a greeting custom, at the all-boy Catholic Christian Brothers Academy in New Jersey, I consistently got astonishingly positive results. We wove many cultural elements into our connection facilitation models. Later, working with at-risk or adjudicated youth, or inner city youth from families with drug addictions, I and the instructors I trained kept having incredibly powerful successes turning these youth's lives around, very quickly, using cultural mentoring based Nature-Connection. We eventually had incredulous social workers and child psychologists coming to find out why our youth were improving, when nothing else had been effective. The authorities told us we must have gotten lucky. Yet, so many of those young people recovered and went on to become high-functioning adults, several even became leaders in their respective fields.

Structural and Pervasive Challenges .

Western society is based on dis-connection and unconscious incompetence when it comes to supporting connection. We've inherited a complex ecology arising from historical trauma resulting in dis-connective behavior patterns that have normalized over time. This normalization creates an unconsciously incompetent crowd mentality upholding dis-connective behavior patterns, learned as children and complicitly replicated by adults. These patterns are bolstered by a form of psychological inertia related to Stockholm syndrome, where captives, over time, come to identify with, empathize with, and vehemently support their captors.[6]

These dis-connective behavior patterns have been passed down through generations for so long that very few modern adults can remember that there ever was connection in the first place. For those of us raised in Western society, there was a moment in history that forever changed our social structures and our system of values. All of our ancestors, who once lived in deep, reverential connection to the natural world, were brutally forced into a system of exploitation and extraction. Whenever an Indigenous, Nature-Connected culture is conquered, usually by the force of expanding imperialism, within a few generations the only thing the remaining people of that culture know is the conqueror's pattern: consume, exploit, extract, and repeat. Western society emerged from that moment when connection was no longer prioritized. What did we lose in the process of westernization? Can we let the cultural patterns of our ancestors inform our modern experience so that we can walk in this world in a more responsible, intentional way?

The life cycle in the dominant Western society, which we could call post-traumatic, results in the loss of the village as a social structure, intergenerational transferred effects of historical trauma, and a worldview emphasizing dis-connection. There is no connection or movement between people in different ages and stages of life. People can end up feeling neglected if they are not welcomed, or angry if they don't get a rite of passage. based on mentoring and a deep sense of connection with nature They can feel isolated if they don't get initiated into adulthood. They can feel hopeless without having a vision. Odawa Elder and Peacemaker, Paul Raphael from Peshawbestown, Michigan offers this view of broken culture at these different ages and stages of life:

1. At birth, neglected
2. Ages 12–13, anger
3. Ages 19–20, isolation
4. Age 25, hopelessness
5. Ages 35–40, bitterness
6. Ages 45–55, emptiness
7. Age 60, loneliness
8. Age 70+, fear

By contrast, the life cycle in a healthy culture involves much connection and movement between various ages and stages of life; villages remain intact and culture is able to regenerate:

1. At birth, welcomed as newborn
2. Ages 12–13, rites of passage, training and preparation for adulthood
3. Ages 19–20, initiation, welcoming into adulthood

4. Age 25, search for personal vision and purpose
5. Ages 35–40, period of deeper service
6. Ages 45–55, community leadership
7. Age 60, teacher and elder
8. Age 70+, spiritual advisor and leader

Paul Raphael told me a healthy culture welcomes newborns. It has Nature-Connection-based rites of passage, training, and preparation for adulthood, and they make sure to initiate everyone into adulthood. People in their twenties receive help searching for a personal vision, then go into deeper service in their thirties and forties. They become real leaders in their fifties, elders in their sixties, and spiritual advisors when they're older than that.

In the healthy culture of the San Bushmen, the Eight Attributes of Connection are present during all ages, from birth until death. They never have a lapse in the attributes, even during the teen years. Their teens have the Eight Attributes, and continue to grow into empowered adults with a clear life purpose. When I saw that, I knew we had the cure for teen suicide right there.

Albert Einstein said, "We can't solve problems by using the same kind of thinking we used when we created them."[7] The complex ecology of historical trauma has generated a kind of thinking that we cannot use to address dis-connection. The current Western framework, based on dis-connective behavior patterns, will not help us find reconciliation or resolution to the issues of historical trauma.

Westernization has been based on a massive dis-connection strategy. After two or three generations, this dis-connection is now a functional part of Western systems and structures. The people who are now stakeholders within the new social fabric have developed a cultural amnesia regarding connection. Western thinking has become reductionist, divided into discrete and dis-connected disciplines of expertise. None has enough perspective to identify or address all of the aspects of the connection problem.

Most modern people don't even know that this is happening because of our consciousness' self-protective defense mechanisms. These mechanisms help people feel both "sane and functional" in a dis-connected society which is neither. People have developed these defense mechanisms to feel safe enough to simply get through the day. Now we know that humans have the potential for a much healthier existence, one that *is* actually sane and functional. But if that isn't modeled for a person, if a person doesn't get to taste that sweetness, they won't know what they're missing (since they can't know what they don't know), and so will continue with their unconscious incompetence in connection.

I propose that we all, as individuals, need to understand the pervasive nature of our cultural amnesia, our complicity in perpetuating historical trauma, and the wide range of socially accepted and normalized dis-connective behavior patterns occurring around us and through us. We need to become expert at our own individual inner tracking, so we can become truly helpful to ourselves and to others.

Paul Raphael, asked, "What do you really long for that was never provided for you?" Then he said, "When you figure that out, become that." For instance, instead of behaving as victims, we can each embody the qualities we were missing. This longing for something that was never provided is essentially a sign that our culture has failed us. As children, we all have intuitive and instinctive longings to allow our nervous system to unfurl like fern fronds out into the world in a gentle and natural way. Often, we are met with harsh moments of external discipline. This can result in feelings of betrayal, reinforcing a dis-connected worldview. We were seeking someone to catch our story, love us, and hold us, but instead we got scolded for a transgression about which we were ignorant. After the initial shock, this pattern repeats multiple times until children begin to lose their trust of adults, or their peers, or their family. Children stop believing in the possibility of connection, love, and affirmation. Nobody wanted to hear about what happened during their time in nature, and this can create a terrible longing that continues being unmet into adulthood. We need a framework of a different scope that shifts the narrative and the conversation so we can see other options. We need another type of modeling. The connection model I propose is aimed at restoring connection, both with nature and with people.

In a German study of forest kindergarten teachers at multiple schools, all the teachers had advanced college degrees, and were considered competent, professional, caring, well-meaning people. The study found that students taught by a Nature-Connected adult became Nature-Connected, while those taught by a dis-connected adult ended up dis-connected. This led to the conclusion that connection modeling is based more on role-modeling than on pedagogy.[8] Educational modeling and connection modeling are good dance partners but they're not the same.

Something was unique about what I was doing in 1983. I was a trained teacher, working at a high school, using vocational-educational modeling. But what I was doing to increase Nature-Connection was completely different, with a whole different set of principles. It was outside of the standard relational models in which I had been trained. Below I list four types of relational modeling:

1. Vocational and Educational Skills Modeling
2. Fitness and Recreational Modeling
3. Connection Modeling
4. Cultural Mentoring: results in Deep Nature-Connection and the Eight Attributes of Connection

Connection Modeling

I found connection modeling to blend effectively and synergistically with both vocational-educational modeling, and fitness-recreation modeling. Compared to their peers, our connected youth were better at sports *and* academic study. They were more focused, more self-driven, and had more creativity. Connection does not distract our young ones from achieving outcomes we say we want for them; it's an enhancement. In professional athletics, processes related to what I have dubbed "connection modeling" are known to be effective. Phil Jackson coached the Chicago Bulls for five years, using connection-modeling-related activities and processes to get them to higher levels of athleticism.[9]

We use certain metrics to measure effectiveness in education, and other metrics for fitness. When we tried to correlate those metrics with connection modeling, we couldn't. We need different metrics for connection, because connection corresponds to a different part of the human nervous system, and so requires different kinds of strategies and support. Intentionally combining the first three of these models together results in the fourth: large-scale cultural modeling, possibly even cultural establishment. I called this Deep Nature-Connection. The metrics of Deep Nature-Connection are the Eight Attributes of Connection (see Figure 11.1, above).

The Model of Connection

We gathered over a thousand different cultural elements and tools that we were testing. And it got very confusing, so we had to create a model. We deliberately based the model on the X, Y, and Z axes for the three spatial dimensions. Later, our traditional Hawaiian friends from Anahola, Kauai added past-present-future as a fourth dimension. It's deliberately not deep or philosophical, or tied to any cultural lineage. It's not a borrowed "medicine wheel" from an Indigenous culture. It's based on primary observations of nature, space, and time. We used universal elements that even modern people can relate to and understand. The model is made up of categories that provide design guidance for applying layers of activities, techniques, and processes. These activities, techniques and processes can be placed on the map of this model based on a number of possible qualitative associations

The 8 Shields

There is one "shield" (collection of processes and techniques) for each of the eight directional archetypes (as related to time of day ... east as sunrise, southeast as

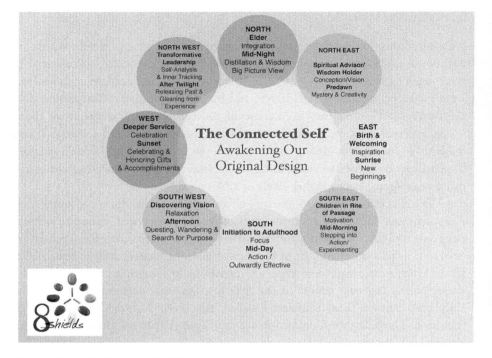

Figure 11.2: The Connected Self.
Source: Author.

midmorning, south as midday and so on). Each of the 8 Shields is designed to achieve a particular outcome in terms of the Eight Attributes of Connection. See Figure 11.2.

This Figure 11.2 depicts a few of the layers that can be used for designing a facilitated connection model experience for participants. The model facilitates safety, since peace-making and community development strategies are at its core. Remember people will not connect unless they feel safe. When we've run programs to increase connection, and the youth haven't felt safe on all levels, they would not connect to anything. They would stay hidden and locked down, since they were so used to programs based on dis-connection. So, when we introduce connection modeling, we need to address that. We use these four peace-making principles and processes that support safety:

1. Restoring healthy connections
2. Collaboration and building consensus
3. Evolving content and creative process
4. Community and sensitive conduct

The Eight Strands

Within each of the eight Shields there are eight Strands, each of which is a pathway of initiation. The Strands repeat on each Shield. Each Strand represents a set of practices and concepts as a progressive pathway that cause an individual or a group to journey towards the center of each Shield. This journey embraces a continuum of experience ranging from a beginner's awareness to the development of an accomplished "master." Strands are comprised of cultural elements that can be applied to create movement from the outer, less experienced, side of the Shields toward the inner, more experienced, center of the Shields.

1. Tasks, village role and leadership role
2. Species and natural elements relationship list
3. Technique related to skills development
4. Attribute of connection
5. Mentoring techniques
6. Cultural elements
7. Regenerative design principles
8. Sensory development

Sixty-Four Cultural Elements

Let's look at one strand named the "Cultural Elements Strand." In the 8 Shields model, this is the Northwest Strand. If we follow the Northwest Strand—on each shield we will see that it crosses 8 Rings. Multiplying this one strand by 8 rings and 8 shields the result is 64 intersections or categories. The 512 Project is currently creating content for each of these sixty-four Cultural Elements. The 512 Project is named after the intersection of all strands, all shields and all rings. There are 8 shields, 8 rings and 8 strands in total ($8 \times 8 \times 8 = 512$).

Figure 11.3 is the depiction of a version of 64 cultural elements used as part of the connection modeling in a program in 2007 for the Regenerative Design and Nature Awareness immersion program held with the Regenerative Design Institute in collaboration with 8 Shields Institute.

The Eight Rings

Each Strand crosses eight Rings on a transformational, healing, empowering, and initiatory journey. A threshold for a rite of passage occurs between the fourth and fifth Ring. The journey from Ring 1 to Ring 8 on each Shield represents a progression of experience and understanding. Each Strand causes movement into deeper

The 64 Cultural Elements for RDNA
February 27, 2007 – JY, Ring 3,4,5
Community Elements that Best Complement Mentoring in Nature

	East	Southeast	South	Southwest	West	Northwest	North	Northeast
1	Greeting Customs	Naming	Taboos	Regalia & Profile	Good Message	Grief Removal Systems	Codes of Respect	Four Directions
2	Moieties	Children's Universal Passions	Peer Mentoring	Imitation	Finding Gifts in Others	Extended Family Ohana lite	Coaching	Humor
3	Nature Museum	Games	Language	Secret Spot	Finding Gifts	Inner Tracking	Honoring Commitments	Story of Day
4	Facilitate Individuality	Journaling	Mapping	Self-Nurturing/ Car-Cleaning	Rite of Passage	Adoption	Honoring of Ancestors	Animal Forms
5	Improvisation & Facilitate Creativity	Mentoring & Art of Questioning	Tracking Culture	Nomadic Culture	Food & feasting	Celebration of Vision	Honor and Unity	Facilitation of Storyteller's Mind
6	Knowledge of Place	Ancestral Tracking	Treaties & Currency of Exchange	Timeless Education	Role Models	Condolence & Mihi Kala	Creative Corrective Discipline	One Liner
7	Vision	Elder Sanctioning	Competition	Song line	Heroes	Sacred Fire Historic Trauma	Oral Tradition	Trickster Transformer
8	Wopila Commitment	Ceremonial Traditions and Sacrifice	Asking Permission and Giving Offerings	Peace of Mind & Big Love	Totems, Affiliations and Medicine Bundle	Intergenerational Transfer & Possession	Specialist and/or Secret Societies	Highly Developed Senses & Aloha

Figure 11.3: The 64 Cultural Elements.
Source: Author.

Rings of experience. People can begin a specific journey from almost any starting point, based on their interests. On the outside of the diagram are beginners, the babies, and in the center are the elders, the 90-year-olds. So, you can say birth is on the outside, death and ancestors are on the inside.

Healthy Culture

Healthy culture predictably leads to connection for members of the culture. Healthy cultures generate, strengthen, tend, and restore connections. Mentoring is a building block of culture. To regenerate properly, healthy cultures require both halves of a circle, one half emphasizing the ecological or Nature-Connection, and the other half emphasizing the social or People-Connection. In a healthy culture, children are mentored in Nature-Connection to increase their awareness. After passing the threshold of adulthood, marked by a rite of passage, initiates are mentored in People-Connection to increase their understanding of ceremony, and support of the family and village.

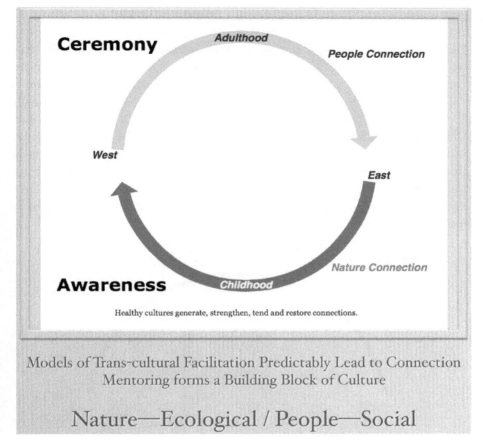

Figure 11.4: Models of Trans-Cultural Facilitation.
Source: Author.

Jon Young's mentor, Tom Brown, Jr., drew the two arrows of Figure 11.4 on a napkin in 1999 to explain a diagram that Tom's elder mentor, Stalking Wolf, had drawn for him in the sand. Tom explained that the top represents "ceremony" and the bottom represents "awareness." Ceremony is only half a wheel and needs to be practiced in the context of nature awareness. Nature awareness needs to be held in the context of the connection that ceremony can provide. Neither awareness or ceremony are complete alone. The two arrows have become an important part of routines of connection.

Most healthy cultures have a day cycle and a night cycle, a Nature-Connection component, and a People-Connection component that are equally weighted. Our work attracts a lot of what I have lovingly termed "sociopathic naturalists": modern adults who fled to nature because of a bad experience with people, and have gotten somewhat Nature-Connected. But increasing Nature-Connection without People-Connection is problematic. Modern adults can only progress in their Nature-Connection journey if they are mentored in People-Connection. That's why we now include People-Connection in our work. I wasn't excited about that, because I started out as a sociopathic naturalist, but I have learned that it's vitally important.

We have also identified many layers in the 8 Shields approach. A commonly applied example of program design might use, for instance, eight categories of nature Connection Routines, eight Concentrations, and eight Physical Senses that overlay and intersect with the archetypal values and stages of life. The development of these Routines and Concentrations results in neurobiologically awakened Physical Senses and the Eight Attributes of Connection. The diagram below (Figure 11.5) shows a few layers of design in a commonly applied connection model pattern utilized by many Deep Nature-Connection practitioners trained in the 8 Shields model.

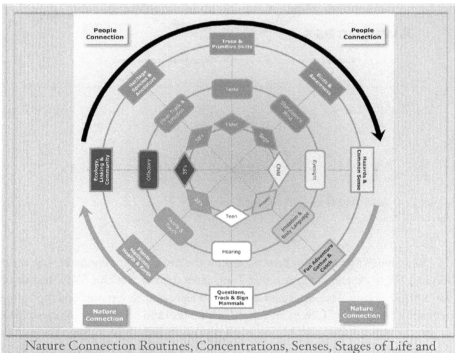

Nature Connection Routines, Concentrations, Senses, Stages of Life and Archetypal Values all intersect and co-create a powerful culture

Resulting in 8 Attributes

Figure 11.5: Intersection of Aspects Resulting in Eight Attributes.
Source: Author.

Deep Nature-Connection, Mentoring, and Culture Repair

We have identified four levels of Nature Relationship

1. Nature Information. Available everywhere via books, media, internet, educational systems and sources.
2. Nature Experience. A multi-billion dollar industry including outdoor recreation, sports, dog-walking, and gardening
3. Nature-Connection. Related to specific demographics (e.g., many baby-boomers), often unintentional and following patterns, while not quite fully activating all of the Eight Attributes of Connection. An important starting place
4. Deep Nature-Connection
   Sign of healthy and effective culture
   Requires strategic Mentoring
   The Eight Attributes of Connection drive transformation
   Results in true transformative eco-social leadership
   Regenerating Deep Nature-Connection strategically

When we attune modern people to the language of nature using applied connection modeling, and their mirror neurons start responding, it can activate connection at a deep level very quickly. Over the last thirty plus years, I have seen this modeling become more efficient and effective through application and iteration. We are seeing the embodiment of the Eight Attributes using a model with predictable and efficient results that is turning the tide of nature deficit disorder.

We can repair our culture and rebuild our communities. Using connection modeling, we can teach skills to people in various ages and stages of life, and we can get cooperation between different groups. Artists, musicians, and elders can contribute as spiritual advisors. Parents with babies can be involved. Parents with young kids in programs can mentor. The model gives us an organizational chart for sorting communities and giving people roles. The model allows us to suggest which tools would be most useful for which groups at which age-appropriate levels. What kind of training parents need, what kind of training grandparents need, and so on. We can then train people in meaningful ways, enhancing connections to other people and to nature.

Nature-Connection will happen naturally because of our neurology, the way our nervous system is wired. But we need strategies to regenerate Deep Nature-Connection. That won't just happen by itself. To achieve that, we have to combine Nature-Connection with educational modeling and outdoor recreational modeling. We need a scale-able model that can be replicated all over the world. And we need more Deeply Nature-Connected mentors to help as many young people as possible.

We're working with Indigenous communities in several locations around the world and giving them back this ancient knowledge so they can use it to self-analyze their communities to see what pieces are missing. Jake and Judy Swamp, and Mike McDonald, from the Haudenosaunee Tree of Peace Society, were so impressed by the connection qualities of the children in our programs, that they asked me to bring this connection model to Akwesasne, where they hoped it would benefit the Mohawk children. We have great relationships with the Hawaiians in Anahola, on Kauai, where the Kanuikapono Public Charter School is effectively applying elements of this connection model and getting amazing results.[10]

We need to ensure that every child in the modern world has the Eight Attributes of Connection. How can we do that? The University of Vermont asked us if we could use an online degree program to get modern adults to embody the Eight Attributes in two years. So I asked, was that even possible? Modern adults are more problematic than the children we'd been working with; children are already much closer to the Eight Attributes. If we gave them the proper conditions, children embodied the Attributes easily. So we did a case study, and we found 125 modern adults who had embodied the Eight Attributes. And 28 of them had done it within two years. Then we asked, what did they all have in common? We hoped that this would be the shortest and most efficient journey to embody the Eight Attributes. We concluded that they all had four things in common, which I call the Four Pillars.

The Four Pillars

1. Deep Nature-Connection Core Routines. e.g., Tracking, Bird Language skills, primitive skills and knowledge
2. Conscious competence in the Art of Mentoring. e.g., Mapping, inner tracking and small group skill.
3. Effective grieving routines
4. Exposure to transformative leadership and role-modeling of the Eight Attributes

The Four Pillars can also be expressed as four layers in an experiential design

1. Skills and Knowledge. Curiosity and competency
2. Recreation and Experience. Good feeling
3. Connection. Instinctive and free-play
4. Deep Nature-Connection. Cultural facilitation

Core Routines

Core Routines of Nature-Connection are detailed in *Coyote's Guide to Connecting with Nature*.[11] These are things that need to be done every day in a regular way. In western education, we sometimes get tired of repeating a basic subject, and want to move on. In Nature-Connection, we have to constantly revisit the same routines to continually deepen our understanding. Some of the most effective core routines are:

- Sit Spot: sitting still in nature, to watch and observe and listen
- Story of the Day: returning from nature to tell your story to somebody
- Storytelling: expanding all your senses to achieve the storyteller's mind
- Bird Language
- Tracking
- Timelessness: to counteract the powerful dis-connection of time sensitivity
- Gratitude: as a daily practice
- Wandering: without a destination, allowing your body to walk you to places
- Sensory Expansion
- Mapping: both verbal and pictorial, for brain development and landscape integration
- Journaling: on your own if no one can catch your story

The Art of Mentoring

When I met Tom Brown, Jr., I was ready to be mentored. His mentoring style, which he learned from Stalking Wolf and his Apache lineage, was so powerful that he brought me to a much deeper level on my Nature-Connection journey. It's important to recognize how effective mentoring can be. I learned the following mentoring questions from Tom Brown, Jr. and Paul Raphael:

- What happened here?
- What's it telling me?
- What's this teaching me on a deeper level?
- How can this help me?
- How can this help me to help others?

In a healthy culture, people receive mentoring at most points during the first half of the life cycle. There is connection and movement between ages and stages of life. Villages remain intact and social leadership structures are able to regenerate.

1. Ages 0–5, supporting parents with young ones
2. Ages 6–12, parents with children in programs

3. Ages 6–12, mentors working with children
4. Ages 13–19, mentors working with rites of passage
5. Ages 13–19, mentors working with young adults on Nature Connection journey
6. Ages 20–25, people searching for their gifts and purpose with Vision Quest related activity
7. Ages 26–45, people stepping into their leadership and modeling connection
8. Ages 46–59, leadership moves into a deeper, mentoring presence, supporting those seeking their gifts and vision; and, general community helpfulness
9. Ages 60–79, elders, the foundation of the community, offering guidance, reflection, support and regenerative wisdom; supporting everyone in the community in a variety of ways
10. Ages 80+, deep and abiding love and spiritual guidance
11. All ages, creatives, storytellers, poets, visionaries, artists, and musicians, all fostering the flow of creativity and movement where needed.

In a healthy, Nature-Connected culture, we find connection modeling elements at every age and stage of life—pre-birth to post death. Before children are born, parents-to-be receive mentoring, even before they marry and conceive. And strong mentoring continues for every child and for their parents. Connection is further enhanced by Mentors who are actively involved in rites of passage and marking thresholds for entering new stages of life.

Grieving

Some modern adults have never grieved, may be ignorant of grieving processes, and are in particular need of grieving. Modern adults whom we've helped to re-connect become sensitive, empathetic and often soon realize all the ways society has failed them. They become hypersensitive to negative ecological and social trends. They realize they didn't get what they needed as children, they start to long for it, and start to grieve. We've learned that we have to help them move through the grief, so they can move forward on their journey to become more connected.

In a healthy culture, grief is tended regularly, almost every day. It is not something done once in a while because somebody dies. The San Bushmen are aware of this, and they have a grieving ceremony four out of seven nights on average.[12] That's a lot of grieving ceremonies. In traditional Hawaiian culture, every family is responsible for having effective grief tending, because each family has daily grief tending needs.[13] If the grief attendant in a family dies, that family would borrow an attendant from another family for a little while. But not indefinitely, because she's got her own family to deal with. So they borrow and train up a new grief attendant to have those skills as quickly as possible.[14]

We need more practitioners who understand Connection-Based Grieving and somatic experiencing, who can bring various deep trauma somatic healing techniques to where they're most needed. We need effective grief healing specialists in a way that we've never needed them before on our planet. Every community needs them abundantly. It's a big job, because we're basically trying to heal thousands of years of historical trauma.

Transformative Leadership and Role-Modeling

We need to train more leaders in Deep Nature-Connection and role-modeling the Eight Attributes of Connection. They can then become those Eagles on the Tree of Life that the Haudenosaunee refer to, those role models who will help restore a healthy culture. And we have a bottleneck. Of the 28 adults who achieved the Eight Attributes in two years, they had two people in common. 25 of them were trained by Gilbert Walking Bull, eight of them were trained by Paul Raphael, and there was some overlap with those who shared both. Gilbert and Paul are both traditional elders who know how to lead traditional healing ceremonies. The problem is, you can't borrow traditional healers from overly taxed Indigenous communities to help everybody else. They've got their hands full.

We need a double strategy to support both leadership and outreach: (1) welcoming and Mentoring newcomers to our Deep Nature-Connection programs; and (2) commitment to long-term mentoring of the alumni of our Deep Nature-Connection programs. I've been working with a wonderfully eclectic global community of early adopters who are really fun. We've been training the leaders of the future, and they are wonderfully inspiring.

Current Projects

The Origins Project

We ask the question, "How can we optimize connection through epigenetic evolutionary potential." We have been working with the Naro San Bushmen in central Botswana who are just incredible people. They are traditional people who don't read or write, who have never known anything but the life of their ancestors, and who have never left the Kalahari desert. And you would think they're unaware of modern things, but they bump into modern people, and they occasionally peek over people's shoulders on Facebook, or get into some cousins' houses and see that they like Jackie Chan films a lot. And they understand what we're working to do through the 8 Shields Institute and Ecotours International. Together with members of the San community that supports our work, we have collaborated to create the Origins Project: to explore the origins of deep nature connection. They're excited that this conversation is happening and they want to contribute. They stand behind us and are incredibly invested in all the ideas we're talking about here. So the Origins Project aims to not only give them self-determination, but give them a job to essentially be themselves. There is a media component to the Origins Project linked to the 512 map, and aiming to document many of the Bushmen's techniques of Nature-Connection at which they excel.

The 512 Project

The model of connection that I am discussing here is the basis for a powerful best-practices database of 512 Cultural Elements at the intersections of the Eight Shields, Eight Strands, and Eight Rings. These cultural elements seem to be universal, as they are practiced by traditional, Indigenous, highly connected cultures on at least three continents. They are not culturally specific traditions. They are a framework, a universal map of the types of tools that create healthy, thriving, and connected culture. Collectively, these tools provide a proven map for healthy, meaningful relationships, group unity, resilience and connection, which can be applied within modern contexts. We can use these tools to effectively empower our social structures, communities, businesses, and families.

We've been collaborating with Indigenous elders in many parts of the world to help us complete the map by telling us about elements we may have missed. And the elders have been taking that map into their communities to restore elements that may have been lost from their original culture.

Other Projects

The Village Builders Network: We've begun training community stakeholders in different regions, introducing much of the information discussed here

Helpers Mentoring Journey: This is for learning the advanced techniques of connection, facilitation of grieving and rites of passage.

Rebuilding Nature Connected Communities through the Art of Mentoring Leadership: Art of Mentoring camps feel like an authentic village where healthy culture has been consciously created, we use these ancestral cultural elements to ergonomically appeal to our bio-neurology, our bodies and nervous systems, which instantly respond by increasing connection

Connection Facilitators Training: A simple, experiential training process that helps people learn the basics of mentoring and developing effective "containers" that support connection

Nature Connection Projects: There are numerous projects around the globe led by people in the 8 Shields network working at varying ages and stages of life, and in a variety of disciplines and applications, we are working on networking more effectively with these innovators through the Village Builders initiative

Bird Language Leaders: Working with 8 Shields alumni and in collaboration with local Audubon Society Centers and other birding enthusiasts, this model is based on cultural mentoring and connection modeling

Conclusion

Our organization has helped to inspire and train leadership for over 400 projects globally that are applying connection modeling effectively, getting the same results in countries all over the world. When I travel, I meet tens of thousands of children who are growing up with the Eight Attributes, who are like little shining lights, and they're all coming from connection modeling. They fill my heart. These children are the reason I do this work.

When we told Jake Swamp about that, he got this big smile on his face and said, "Hey, this work is taking root." He was beaming with positivity. Then he said, "You know it's going to take about 200 years." And we all drooped to hear that. And he said, "It's important you know that it's going to take 200 years, because then you start thinking of yourselves as foundation builders instead of roofers. You all want to finish the roof and move in tomorrow. There is no village now. You want it now, you want your longhouse complete, but you might not get that in your lifetime. So do you want your children to have that? Do you want their children to have it? Think seven generations into the future. That's your job: to rebuild a beautiful place for us to all dwell together. You're the foundation builders. Ask yourselves: what do the foundation builders need to do?" So that's who we are. We're building the foundation for the future.

Discussion Questions

What is meant by Nature-Connection?
What are the Eight Attributes of Connection?

How is "practiced immersion in nature" something foundational to human nature for learning and development?

What is meant by grieving for the earth in this chapter?

What is Connection-Based Grieving?

Notes

1. Richard Louv, *Last child in the woods: Saving our children from nature deficit disorder* (Chapel Hill, NC; Algonquin, 2005).

2. Darcia Narvaez, *Neurobiology and the development of human morality: Evolution, culture and wisdom* (New York, NY: W. W. Norton, 2014).

3. Roianne R. Ahn et al., "Prevalence of parents' perceptions of sensory processing disorders among kindergarten children," *American Journal of Occupational Therapy* 58, no. 3 (2004): 287–93, accessed online 08 September 2018, https://www.spdstar.org/sites/default/files/publications/prevalenceofparentperceptionsofSPDamongkindergartners.pdf. See also Barry E. Stein et al., "Postnatal Experiences Influence How the Brain Integrates Information from Different Senses," *Frontiers in Integrative Neuroscience* 3 (2009): 21, http://doi.org/10.3389/neuro.07.021.2009.

4. Gibert Walking Bull, personal communication to the author.

5. Brenna M. Henn et al., "Hunter-gatherer genomic diversity suggests a southern African origin for modern humans," *Proceedings of the National Academy of Sciences*, 108, no. 13 (2011): 5154–62, accessed online 08 September 2018, https://experts.umich.edu/en/publications/hunter-gatherer-genomic-diversity-suggests-a-southern-african-ori.

6. Thomas Strentz, "The Stockholm Syndrome: Law Enforcement Policy and Ego Defenses of the Hostage," *Annals of the New York Academy of Sciences* 347 (1980): 137–50.

7. Albert Einstein, but also paraphrased by Ram Dass: "The world that we have made as a result of the level of thinking we have done thus far creates problems that we cannot solve at the same level as the level we created them at." Albert Einstein, "The Question of Consciousness," *The Journal of Transpersonal Psychology* 1, no. 4 (1969):124.

8. Peter Häfner, *Natur- und Waldkindergärten in Deutschland—eine Alternative zum Regelkindergarten in der vorschulischen Erziehung* [Nature and forest kindergartens in Germany—an alternative to regular kindergarten in preschool education], (Ph.D. diss., University of Heidelberg, 2002), accessed online 08 September 2018, http://archiv.ub.uniheidelberg.de/volltextserver/3135/1/Doktorarbeit_Peter_Haefner.pdf; Sara Knight, *International Perspectives on Forest School: Natural Spaces* (London, UK; Sage, 2013).

9. Phil Jackson and Hugh Delehanty, *Eleven Rings: The Soul of Success* (New York, NY: Penguin, 2013).

10. See www.kanuikapono-charter-school.org.

11. Jon Young, Ellen Haas, and Evan McGown, *Coyote's Guide to Connecting with Nature* (Santa Cruz, CA: Owlink Media, 2010).

12. Christopher Low, *Khoisan Healing: Understandings, Ideas and Practices* (Ph.D. diss., University of Oxford, 2004), accessed online 08 September 2018, http://thinkingthreads.com/files/Khoisan_thesis.pdf.

13. Jessica Garrity, *"Hawaiian Grief Culture,"* lecture 2015, accessed online 08 September 2018, https://prezi.com/loaz-qvniwa2/hawaiian-grief-culture/.
14. Donna E. Hurdle, "Native Hawaiian Traditional Healing: Culturally Based Interventions for Social Work Practice," *Social Work* 47, no. 2 (2002): 183–92, accessed online 08 September 2018, http://sw.oxfordjournals.org/content/47/2/183.full.pdf.

References

Ahn, Roianne R., Lucy Jane Miller, Sharon Milberger, and Daniel N. McIntosh. "Prevalence of Parents' Perceptions of Sensory Processing Disorders Among Kindergarten Children." *American Journal of Occupational Therapy* 58, no. 3 (2004): 287–293.

Einstein, Albert. "The Question of Consciousness." *The Journal of Transpersonal Psychology* 1, no. 4 (1969):124.

Garrity, Jessica. *"Hawaiian Grief Culture,"* lecture 2015. Accessed online 08 September 2018. https://prezi.com/loaz-qvniwa2/hawaiian-grief-culture/.

Häfner, Peter. *Natur- und Waldkindergärten in Deutschland—eine Alternative zum Regelkindergarten in der vorschulischen Erziehung.* Ph.D. diss., University of Heidelberg, 2002. Accessed online 08 September 2018. http://archiv.ub.uniheidelberg.de/volltextserver/3135/1/Doktorarbeit_Peter_Haefner.pdf

Henn, Brenna M., Christopher R. Gignoux, Matthew Jobin, Julie M. Granka, J. M. Macpherson, Jeffrey M. Kidd, Laura Rodríguez-Botigué, et al. "Hunter-gatherer genomic diversity suggests a southern African origin for modern humans." *Proceedings of the National Academy of Sciences*, 108, no. 13 (2011): 5154–62.

Hurdle, Donna E. "Native Hawaiian Traditional Healing: Culturally Based Interventions for Social Work Practice." *Social Work* 47, no. 2 (2002): 183–92.

Jackson, Phil, and Hugh Delehanty. *Eleven Rings: The Soul of Success.* New York, NY: Penguin, 2013.

Knight, Sara. *International Perspectives on Forest School: Natural Spaces.* London, UK: Sage, 2013.

Louv, Richard. *Last Child in the Woods: Saving Our Children from Nature Deficit Disorder.* Chapel Hill, NC: Algonquin, 2005.

Low, Christopher. *Khoisan Healing: Understandings, Ideas and Practices.* Ph.D. diss., University of Oxford, 2004. Accessed online 08 September 2018, http://thinkingthreads.com/files/Khoisan_thesis.pdf.

Narvaez, Darcia. *Neurobiology and the Development of Human Morality: Evolution, Culture and Wisdom.* New York, NY: W. W. Norton, 2014.

Stein, Barry E., Thomas J. Perrault Jr., Terrence R. Stanford, and Benjamin A. Rowland. "Postnatal Experiences Influence How the Brain Integrates Information from Different Senses." *Frontiers in Integrative Neuroscience* 3 (2009): 21. http://doi.org/10.3389/neuro.07.021.2009.

Strentz, Thomas. "The Stockholm Syndrome: Law Enforcement Policy and Ego Defenses of the Hostage." *Annals of the New York Academy of Sciences* 347 (1980): 137–50.

Young, Jon, Ellen Haas, and Evan McGown. *Coyote's Guide to Connecting with Nature.* Santa Cruz, CA: Owlink Media, 2010.

CHAPTER TWELVE

Wisdom, Sustainability, Dignity, and the Intellectual Shaman

SANDRA WADDOCK

Ancient wisdom traditions of shamanism could, if applied in the modern world, be helpful in guiding our world towards sustainability. That is because at its core shamanism is about healing man's relationships, including a connection to nature that helps to bring about healing or wellbeing for all aspects of nature, including humans and their systems.[1] The shaman is, according to Igor Gorkov, a lover of knowledge. In the Tungus language from which the word is derived, "sham" means knowledge, and "man" means liker, therefore, lover of knowledge.[2] Shamanism is fundamentally using "nonordinary" experiences, like sudden insights, intuitions, and knowings, to heal relationships by bringing balance, harmony, healing, connections, sensemaking (including healed mythologies), and constructive new memes to our world.[3] Shamanism's purview includes humans, of course, but also extends to the rest of the natural world, and even the planet as a whole.

The core idea of this chapter is that many more of us can—and need to—exert our own shamanic power in our modern world to help heal the troubles that the planet is facing. To do so, we must be willing to take the necessary risks and follow what amounts to a calling to become "fully who we must be." The intellectual shaman addresses issues holistically, and challenges existing theories, ideas, and practices where they need to be challenged as a means of changing the core memes and cultural myths that surround our societies. Thus, as shamans we need to orient ourselves to healing what is within our purview to heal. For scholars, that tends to be theories, ideas, practices, or the memes (core cultural artifacts) that shape our perspectives on the world. Further, I believe that many more of us *need* to undertake these shamanic tasks in whatever walk of life we are in, because our world is desperately in need of healing. For academics in any field that means becoming intellectual shamans.[4] For others, it simply means finding and using your own shamanic gifts.

Traditional shamans, that is, medicine men or women, spiritual leaders, and healers, can be found in virtually all traditional and Indigenous cultures of the world.[5] Importantly, they are also, though often unrecognized, present in modern cultures. The shaman's work emphasizes bringing about harmony and healing in all sorts of relationships, between spirit and self, self and others, humans and nature.[6] With all due respect for traditional cultures' shamanistic customs, it is my contention that the modern world, fraught with problems as it is, is much in need of the healing that shamans adapted to our times can potentially provide. Shamanistic perspectives can potentially help heal the fractures that modernist perspectives have brought to how businesses and economies act today.

The Intellectual Shaman

Intellectual shamanism can be defined as intellectual work that emphasizes healing, connecting, and sensemaking in the service of a better world.[7] In traditional cultures, the shaman is a revered community figure, the wise man or woman whose knowledge helps both individual patients and the community heal when there are problems, dis-ease, or dis-order that need to be corrected. Traditional shamans do much of their healing work by accessing spiritual realms to seek and bring back information that can help heal the patient—and the surrounding cultural mythology.[8]

In crossing from the everyday to spiritual realms, or what Frost & Egri call boundary spanning or mediating realities, shamans are making connections that others do not readily make.[9] In bridging realms—that is in using the connecting function of the shaman—shamans bring new (healing) information back to everyday life in the hope that the shaman can provide a better story or cultural mythology that can help to heal the patient or community.[10] In many traditional cultures, shamans believe that the reason patients become ill—dis-eased or dis-ordered—is that the community's cultural mythology has become dis-eased or dis-ordered, i.e., that the relationships among entities in that community are somehow out of ease or order. To heal the patient, order and ease must be restored in the relevant cultural myths. A new story (or stories) must be told, and the information gathered in other realms helps inform those new stories. Shamans accomplish this restoration, clarification, and reframing of cultural mythology through their sensemaking function, which Frost & Egri claim is a function of the spiritual leader.[11]

In most modern, particularly Western, cultures, the shaman is little understood and, except in some spiritual circles, less accepted. Yet the shaman's three tasks of healing, connecting, and sensemaking in the interests of healing, whether it is the patient, the community, or, indeed, the world, still need to be performed. Shamans can be found in virtually all walks of life, and often include spiritual leaders, artists, business, political, and civic leaders, explicit healers like doctors, nurses, psychologists, and therapists of many types, and teachers of all sorts. The general ideas here can be applied broadly. Intellectual shamanism is reflected in the intellectual, thinking, and writing work of connecting ideas, insights, and theories, emphasizing new ways of practicing, e.g., management and business.

In what follows, I will try to illustrate how a more "connected" shamanistic perspective can help intellectuals and others shape a new worldview around businesses, economies, and sustainability practices. I believe that a shamanic perspective can reshape core memes that influence the stories that we tell ourselves about, for example, why businesses and economies exist and how they operate in the world. Memes are basic cultural artifacts—ideas, images, concepts, phrases, for example—that are building blocks of cultures, ideologies, and, generally, belief systems. The term meme was coined by Dawkins as an analogy to the work of the gene in biology.[12] Memes, as Blackmore has argued, shape culture, and hence, though too often unrecognized, are the foundations of attitudes, beliefs, and perspectives that shape the stories we tell ourselves about how the world operates.[13]

Obviously, this task is huge. The discussion here can only begin what is likely to be a lengthy and difficult conversation in a world fraught with complex challenges. The healing, connecting, and sensemaking roles of the shaman in healing our societies, however, have never been more important. This chapter focuses first on a very brief explication of why shamans as healers are needed more than ever, then elaborates the central role of dignity in a modern shamanic perspective, and finally applies these ideas about dignity to sustainable enterprise and economy.

Why Shamans? A World in Trouble

The world today needs shamans more than ever to help us all heal the fractures in community, in ecology, and in the growing inequity that is driving ever-greater divides between the very well off and the rest of humanity, i.e., the vast majority of people the Occupy movement called "the 99%." The Intergovernmental Panel on Climate Change (IPCC) has argued that the planet faces a human-induced sustainability crisis of monumental and some say potentially catastrophic proportions.[14] Gilding, for one, argues that it may well take a cataclysmic ecological or systemic collapse of some sort for humans to finally get enough initiative to truly act on climate change. Despite the fact that 97–98% of climate scientists now agree that climate change has anthropogenic (human) roots (while recognizing that some amount of climate change is always occurring), such change has been slow in coming.

Further, inequality is increasing both in the United States, where the very rich are capturing ever greater proportions of wealth gains, and globally. More disturbingly, as Piketty has convincingly demonstrated, the current system of economics is likely to lead to the rich becoming richer, while the poor(er) stagnate.[15] In the United States in 2015, the Pew Research Center found the middle class to be shrinking, making the Occupy Movement's 99% and 1% construct even more of a reality.[16] These data indicate that the 1% have captured ever-more of the nation's wealth, while globally the middle class is more of an aspiration than a reality.[17] Of course there are many other problems in the world, but there is a reason that these two trends are significant. Jared Diamond has compellingly demonstrated that civilizational collapse occurs as a result of one of two dynamics—growing gaps between the rich and poor, and overuse of ecological resources, resulting a loss of ecological sustainability.[18] Growing evidence suggests that both of these troubling factors may be at play in today's world.

Given these trends and realities, there may well be an imperative to reverse these dynamics if humanity is to avoid an existential crisis. A modern shamanic orientation to the world around us could potentially be foundational in shaping a new worldview encompassing economics, business, and sustainability.

Dignity and the Shaman

At the core of the shamanic perspective proffered here is the idea of according dignity—intrinsic worth—not just of humans, but also to other living creatures and aspects of nature. Shamans and many Indigenous peoples believe that there is Spirit in everything.[19] That includes human beings, of course, but also other living creatures, as well as aspects of nature that we do not typically consider to be "alive." That perspective also suggests that Lovelock's notion of the Earth as Gaia, a living entity, is highly relevant to a shamanic perspective. Basically, the belief that everything surrounding us—nature and its components, as well as other human beings—is imbued with Spirit, if widely adopted, would mean that we would have to expand the idea of dignity—which is typically only applied to human beings—to other aspects of nature.[20] But we need not necessarily go to the spiritual realm to believe that there is merit in the idea of according dignity to all of Earth's living beings, its natural elements, and the whole of the planet. That perspective can be adopted simply by recognizing our fundamental and inextricable interconnect-edness with the core of energy that manifests in all aspects of the world and that physicists now claim is the basis of reality.[21]

What if such a perspective around interconnectedness and interdependence with all of Earth (and, indeed, the Universe itself) became an imperative of our economic, business, and sustainability thinking? That imperative demands that we humans treat the world around us and all other beings *as if* they had inherent dignity. Let us explore in a bit more detail what dignity means and then turn to some of the implications of such a shamanic perspective brought broadly to bear on our major economic and perhaps political communities, organizations, and systems.

"Dignity" typically means that a person has intrinsic worth—the quality of being worthy or honored, just because they exist.[22] That means that we need to accord dignity to other humans just because they are, not because of what they have accomplished, who they are, or what they own. Dignity applied to human beings means that everyone, no matter what their station in life, should be treated as an end in, rather than merely as a means to someone's or some entity's (e.g., a business's) ends, as many philosophers argue. Notice that adopting such a (shamanic) perspective is consistent with much thinking on ethics and would change the way many people treat others. Each person would be valued for him or herself, not for what they can do for someone or for what they have or have done. Such a perspective would shift the exploitation of people that often takes place in businesses, where people are paid the lowest possible wage, where companies search out the low-cost places in the world, where they can be "efficient" by exploiting human needs for earning a living and supporting their families—then pay workers wages that do not allow them to live with dignity.

Hicks differentiates dignity from respect.[23] Dignity, she argues, is inherent to all of us, simply because we exist. Dignity is integral to us as humans. Respect, on the other hand, must be earned. If we take the shamanic perspective and view others as having Spirit, then this perspective focuses us, almost automatically, on according dignity to others—and respect to those whose actions, behaviors, ideas, or whatever have earned them our respect. Respect means having deep admiration for something or someone. Thus, respect tends to derive from our feelings about others' abilities, achievements, qualities, or accomplishments, or perhaps because they offer wisdom, knowledge, love, or something else that leads us to respect them. But no one needs to "do" or "be" something to be accorded dignity, because dignity is simply integral to human existence.

According to Hicks, we humans are all worthy of retaining our dignity, including being able to make our way in the world successfully.[24] Sen argues from a similar point of view that humans need to be able to use their capabilities to ensure adequate "functionings."[25] That is, they should be able to tap enough resources to live fulfilling lives and to attain wellbeing. If we see dignity as inherent to all others, then there is significantly less chance that we will mistreat them—or design economic, political, business, or other systems and institutions that mistreat others.

Now if the idea of dignity is expanded to encompass other living beings and indeed all of nature itself, then the dominantly Western idea that nature is only "there" for human exploitation would also have to shift. Humankind would take its (rightful in my view) place as *part* of nature, but not necessarily as the central or "highest" (and most worthy of dignity) part. The reality is that humans are fundamentally and inextricably interconnected with nature.[26] We humans draw all the resources that we need to survive from nature. We simply cannot continue to exist, or experience wellbeing, unless nature's resources can support both the number of humans on the planet as well as the ways in which those humans interact with nature.

From this perspective of the shaman who believes that all manifestations of nature have Spirit, all living beings and even "nonliving" aspects of nature have inherent worth, simply because they exist.[27] That means, for example, that rivers, mountains, lakes, plains, oceans, forests, and other elements of nature have Spirit—and should be treated respectfully—as if they had inherent worth and dignity in themselves. Much the same could be said for Earth or what Lovelock calls Gaia— as a whole, and whose ability to support human existence Lovelock believes is threatened.[28]

Dignity violations are hurtful and need to be taken seriously, wherever they occur. As Hicks points out, dignity violations appear to be at the core of almost all conflicts, local and global.[29] She notes that when such violations take place, it is important that they be acknowledged and, to the extent possible, corrected so that healing can take place. This distinctly shamanic perspective, applied to natural elements, would suggest that very different approaches to business and economics might be needed if the world around us is to be according dignity. Such a mindset is radically different from the typical Western mindset that views nature as "there" for mankind, to be exploited. It would imply a respectful relationship with all elements of nature on the part of businesses and other human institutions, away from exploitation and efficiency and growth at all costs. Instead, such a shamanic mindset, honoring all of nature, all of humans, all living and non-living entities, would focus businesses on healing their relationships with humans and nature in ways that create balance, harmony, and resilience in all. From this shamanic perspective on dignity, we move now to consideration of sustainability as part of a shamanic approach to business.

Sustainability, Collective Value, and Business: A Shamanic Perspective

The notion of sustainability has gained traction over the past several decades as humanity continues to push the limits of nature's ability to support the human project. In 1987, the Brundtland Commission introduced the idea of sustainable development and sustainability in its report called *Our Common Future*. The report defined sustainable development as "development that meets the needs of the present without compromising the ability of future generations to meet their own needs."[30] The idea of sustainability emphasizes living within one's means at the macro scale and, when linked to dignity, could include the Lakota peoples' idea, which is broadly embraced by Native Americans and was codified in the Iroquois Great Law of Peace, that all decisions should be made keeping the "seventh generation" out in mind.[31]

Imagine for a moment that the shamanic ideas about dignity are actually applied in businesses and economies. Rather than a growth-at-all-costs mentality, businesses and economies would foster a resilience economy, based in diversity and selective use of natural resources that would be honored in their use, much as Native Americans honored the spirits of the animals they hunted. Rather than exploiting nature by, for example, blowing off mountaintops to get to the resources inside then dumping the toxic residue in nearby valleys, clear cutting forests and thereby destroying not only animal habitat but also the diversity of the natural forest, or dumping toxic wastes into rivers, streams, and lakes, we would accord dignity to these natural elements. Note that such a stance does not mean not using resources: it means treating natural resources as sacred and using them accordingly. Fracking, which harmfully removes difficult energy resources from the environment, and shipping oil through long pipelines across pristine territories, digging for oil in still pristine areas, and many other destructive business practices would be a thing of the past. No longer could businesses simply focus on exploiting natural resources, rent-seeking, growth and "efficiency" no matter the human or ecological costs. Businesses would have to refocus on restoring what was removed, removing no more than nature (or man) could restore, and treating all natural manifestations as sacred.

A shamanic perspective would thus bring a very different attitude towards people, nature, and all living beings, because all of nature would be accorded dignity—worth—simply because it exists. Of course, that does not mean that change is not possible and even necessary. Rather it means that natural manifesta-tions and other creatures need to be treated well, despite the fact that sometimes we will be, e.g., using land and natural resources, eating animals, and building various constructions for human use. We humans are, after all, as I have argued in an earlier paper, all stakeholders to Gaia, our Earth.[32]

By extending the idea of dignity to Earth itself and all of its manifestations, notions of business value creation shift rather dramatically, and perhaps allow the type of new theory of business, that for which scholars Thomas Donaldson and James Walsh have argued, to emerge.[33] Donaldson & Walsh argue that the central function of business, in their new theory, would be to create what they term collec-tive value (for all, which extends to the Earth and its many manifestations).[34] They define collective value as "the agglomeration of Business Participants' Benefits, ... net of any aversive Business outcomes [caps in original]," with the important caveat that no dignity violations occur.[35]

Building a distinctly normative perspective, Donaldson & Walsh forcefully argue that the dignity criterion provides what they term a "moral 'bright line rule.'" That moral rule provides guidance as to what types of business activities are pro-hibited, including, of course, violations of human dignity, including upholding basic human rights.[36] In the spirit of sustainable development, these authors fur-ther argue that business decisions need to be accountable to past and future, as well as to present generations because "legacy matters."[37] As Donaldson & Walsh state, "We owe it to our ancestors, ourselves, and to our descendants to strive for optimized collective value. Work for anything less and we will very likely get it."[38] Although they do not develop the idea fully, Donaldson & Walsh also implicitly reference what I have here called the shamanic perspective of dignity in natural manifestations, stating,

> While we defined dignity in terms of the inherent worth of persons, we now wonder
> if non-human animals manifest any kind of dignity. Indeed, might other parts of the
> biosphere, forests, grassland and coral reefs, possess their own form of inherent worth?
> We think the answer to both questions is yes, although we acknowledge the difficulty
> of justifying this "yes."[39]

The shamanic perspective provides a way to answer these questions affirma-tively even if one is only acting "as if" there were inherent value in all of nature's manifestations.

Humans, Dignity, and Sustainability

Perhaps most importantly, a shamanic perspective that accorded dignity to all people regardless of their status in the world might well upend many of today's "management," "efficiency," and growth-oriented practices that have harmful social or ecological consequences. If businesses actually treated all people with dignity, for example, then workplace conditions, pay scales, and environmental practices in developing nations would dramatically improve. Such practices would also be in accord in industrialized and developing nations with the United Nation's Universal Declaration of Human Rights, passed in 1948. No more could workers be dehumanized by their working conditions. They would need to be paid a living wage—a wage that permitted them to live reasonably well, eat enough nutritious food to sustain their health and wellbeing, and support their families adequately. Decent work, as the International Labor Organization calls it, and decent wages would become the norm, rather than the near-slave pay, excessive working hours, and substandard conditions that now exist in many global supply chains, where companies are seeking the lowest cost wages and working conditions. Under a shamanic perspective, workers would never have to live in poverty conditions and should, in such conditions, be able to work their way towards standards of living that enable dignified lives for all.

Developed nations would also be subject to the same higher standards for their workers. Dehumanizing and demeaning tasks that provide little useful benefit and simply foster ever-greater growth and dehumanized conditions (e.g., call centers) would have to be rethought. Factory conditions around the world would need reconsideration so as to avoid the dehumanizing practices rampant in so many places. The employer-worker relationship would be imbued with dignity-enhancing relationships, with all workers valued for the contributions they make, and the meaning and contribution of their work to the greater good kept firmly in mind.

Nature, Dignity, and Sustainability

In a collective value and dignity-based orientation, businesses would work in harmony with nature, drawing out resources at rates in which they can be replaced, ensuring that the diversity of natural resources remained and that resilience be sustained. Resilience and diversity of nature would be central goals because of their centrality to fostering wellbeing for creatures and the land, as well as their restorative powers.[40] Farming and agricultural production practices would shift to honor the dignity of the animals, even when they were destined for slaughter and ultimately consumption. The inhumane conditions in which many animals are raised today would be replaced by significantly more humane conditions, where animals were free to roam in natural habitats and allowed to live dignified lives.[41] Slaughtering animals might need to be treated as a sacred event—not as mass production—and, indeed, how and when to eat meat might also be rethought, especially given the significant negative externalities associated with meat production and consumption.

Treating animals as if they had Spirit, i.e., recognizing their sentience and awareness, would thus mean that animal husbandry moves away from the industrial practices prevalent in many places today. No longer could animals be kept confined in spaces too small for them to move, wallowing in their own excrement, and fed corn or other products that do not maintain their health.[42] No longer would antibiotics be delivered to animals living in unhealthful and close conditions to keep them "healthy." As a result, the side-effects that antibiotic overuse has caused, including increasingly drug-resistant bacteria and antibiotics in rivers and streams, among other negatives of the current production system, would reduce.[43]

In such modified approaches, animal waste would be naturally processed (and improve the land, rather than polluting it), rather than "processed" into giant waste pools that pollute nearby streams and rivers. If field and crop rotation were used dynamically, as opposed to today's monoculture crop production and industrial animal husbandry practices, animal waste would become food for other living beings. Such practices could enrich soil that could then be used for planting, rather than having that soil be constantly depleted by industrial farming practices year after year.[44] Organic and biodynamic production methods would replace industrially-oriented farming practices, such as excessive use of pesticides and fertilizers. Potentially lower "productivity" rates for both crops and animals achieved through such methods would mean that more jobs, which are much-needed in many parts of the world, would be available on nature-based farms. Further, many more people might learn how to grow their own crops and at least partially sustain their own lifestyles. Probably not incidentally, food products would become inherently more nutritious.[45] This approach is not to advocate a return to old ways, but rather to apply modern understandings about sustainable and dignity-based agricultural approaches to animal husbandry, crop production, and general land (and forest) usage.

Treating the land, oceans, rivers, and lakes as if they had dignity would mean considerably different practices for many businesses. If the shamanic perspective were adopted, then lakes, rivers, streams, and oceans would be treated like living entities. Polluting them with runoff from industry and agriculture would become considerably less appealing, as would polluting the atmosphere. If we humans viewed mountains as sacred, then the likelihood that their tops would be blown off to get at the resources inside them would diminish. There would also be no such thing as mine dumps, which are toxic enough that little grows on them, that now desecrate too many valleys and open spaces.

Greater marine stewardship, which is implied in a dignity-based approach, means that practices like ocean trawling (in which giant nets scrape up everything in their paths, leaving devastated ocean bottoms), would need to cease. They could be replaced by less "efficient" (but more effective, in Drucker's sense) practices that selectively gather marine resources for human use, leaving enough oceanic life that replacement becomes possible and the diversity of the food chain that produces larger ocean creatures is maintained.[46]

Much the same could be said of food production processes, which would have to move away from artificial fertilizers, pesticides, and genetic engineering that pose significant risks, not only in their misuse of land, but also in their potential for pollution, and in their generated uncertainties about long-term safety. Similarly, food production itself would need to focus on honoring the ingredients that go into food—moving away from excess processing of food "products" that, while seemingly convenient, actually have little nutritional value and contain high amounts of salt, sugar, and fat. Some of these "products" are designed to get people to eat more than is healthful, playing a significant (and largely unspoken) role in the obesity crisis that developed nations are facing.[47]

Creating Circularity

Many will undoubtedly argue that such a shamanic perspective on businesses and industry is unrealistic in our modern era, or that it would seek to return us to an imagined and idealistic past. But to the extent that current practices in businesses, agriculture, marine production, and economics generally have created an unsustainable world fraught with dignity violations, those practices need to be rethought. By recognizing the very real problems that our linear production processes (i.e., raw-material to factory to store to home to landfill) have created, perhaps it is not too "far out" to consider a more shamanic (and nature-based) approach. Such an approach might have the potential to bring more creativity, more productive and meaningful work, more dignity for all, and a "sufficient" amount of goods and services to support human civilization. Some call this approach a circular economy, where nothing goes to waste, while others call it sustainability or sustainable development.[48] A more shamanic approach to business and economics would keep collective value, wellbeing for all, dignity, and sustainability firmly in mind. For example, a global coalition of interested thought leaders called WE-All, the Wellbeing Economy Alliance, is creating a new narrative about the role of business in society that can potentially help reshape the business and economic mindset in ways that can help us deal with the uncomfortable new realities of our world.[49]

Such a re-visioning of the relationship of businesses and economies to people and other aspects of nature brings us back to the idea of collective value, broadened to encompass all living beings, and all of nature in the shamanic sense. This broad understanding of collective value means value for all, not just business value, but societal and, in a very real sense, ecological value. It emphasizes the wellbeing of all, including other living beings and nature itself, rather than just maximized profits for a group of shareholders. For example, the Organisation for Economic Co-operation and Development (OECD) has defined wellbeing as comprising "two important domains: material living conditions and quality of life."[50] The OECD's "Better Living Index" creates a multi-dimensional approach to wellbeing that emphasizes people, goes "beyond GDP [gross domestic product]'s" known limitations, considers how wellbeing is distributed, and takes into account future as well as current wellbeing.[51] The Better Living Index takes eleven factors into account: income and wealth, jobs and earnings, housing, health, education, work-life balance, civic engagement, social connections, environmental conditions, personal security, and subjective wellbeing.[52] It also addresses the sustainability of wellbeing by measuring four forms of capital: natural, economic, human, and social, providing a holistic way to address wellbeing at scale.[53]

The shamanic perspective would dramatically shift ideas about when, why, and how many goods and services are needed and, most importantly, about how they are produced. The focus would be on the general wellbeing, including that of nature and natural elements, and, of course, human wellbeing, not simply the economic activities that are measured by gross national product (GDP), which has been known to be a flawed measure since its inception.

Businesses in such a system would be asked to treat everyone and nature with dignity and to generate collective value (as discussed earlier and exemplified in the OECD's Better Living Index), not just shareholder wealth. People would need to be treated in dignified ways, no matter what their place in the system. Nature's resources would be honored, rather than exploited. Modern techniques and approaches could be integrated with traditional ways of honoring the earth so that sustainability was at the core of production practices. Generally, the focus of business would shift away from growth-at-all-costs and "efficiency" that does damage towards nature and her creatures solely for the sake of profits. The shift would be towards providing goods and services that are truly necessary, that are durable and of high quality, and that are fully priced out, including all of what are now considered "externalities" or negative by-products of doing business. While efficiency would remain important, it would not be "at all costs" to people or nature. New methods, or perhaps adapted old methods, would need to be developed that treat the land, the topsoil, natural resources, and all other living creatures with dignity.

As I have elsewhere argued, principles/values of renewal, regeneration, and resilience would dominate over other values.[54]

More organic (and biodynamic) approaches to land cultivation and agriculture would create farming practices more in harmony with nature, and would require significantly less artificial fertilizer, pesticides, and mono-cultural practices than do today's industrial agriculture approaches. Animals would be treated more humanly if raised on actual farms as part of the rotation of land use, potentially without significant loss of productivity—and possibly with more generation of much-needed employment opportunities for many.[55] Product design, innovation, and development might need to mimic nature in what Benyus calls "biomimicry," to reduce ecological costs and ensure much more harmony with natural processes.[56] Production would be geared to sufficiency—enough—rather than constant growth, and pricing would have to reflect the reality of the costs involved, including a life-cycle analysis and externalities.

Such approaches might also provide new forms of employment that respect both people and planet. That path of development would enhance human dignity because putting wellbeing, not profit, at the center of business activity would mean less focus on growth and profitability and more focus on generativity, dignified employment, useful and durable products, among other benefits. The focus in a shamanic perspective would not be on constantly speeding up to get "more" of whatever is being produced or purchased, but rather of enjoying what is, in this moment. "Connecting," one of the core shamanic functions, could be done among people face-to-face as well as online, and the healing that human expression and interactions can bring in numerous forms of artistry might well be more valued than it is in today's world.

Although difficult decisions would have to be made, these approaches would also help to deal with issues of inequality. Resources would be more evenly distributed and more people would be employed in productive work, while still allowing for some of the benefits that capitalistic approaches offer in terms of incentives for sound use of resources. What Homer-Dixon called the "ingenuity gap" might be solved because, as Gilding points out, human beings rise to the occasion in terms of ingenuity, innovation, and creative enterprise when a crisis—or major change—is upon them.[57]

Prices for products produced anywhere in the world would need to reflect these changes. From a sustainability perspective—both for the company and the planet—that might not be as bad a thing as it initially seems. Higher prices would foster decreased consumption—and consumption and constant "growth" (ever-increasing sales, materialism, and consumption) is hardly aligned with the planet's capacity to support human civilization. Ehrenfeld & Hoffman and Jackson, among others, have clearly demonstrated that infinite growth is simply not feasible on a finite planet.[58] Thus, a collective value and wellbeing approach to business and economics demands a more resilience, diversity-based, and ecologically realistic approach. Higher prices, resulting from more humane and dignified treatment of workers (and, as below, animals and the land itself), would result in less consumption and, most likely, higher prices as externalities are incorporated. Customers, facing these higher costs, would demand better quality for the products that they were able to purchase—and better quality would result in greater durability. Combined with excellent design and styling, quality could foster better customer satisfaction with many products, while substantially reducing the ecological resources needed to produce them.

In other words, a collective value and dignity-based approach to business would foster an entirely different business, consumer, and social mindset focused on wellbeing, quality of goods and services, durability, and "effective" use of resources. Effective use of resources moves away from the "efficiency"-at-all-costs mentality that now drives many businesses. Efficiency has (not always intentionally) meant producing and attempting to sell as many products as possible, creating an ever-changing revolving set of pseudo-innovations that drive constant replacement of many still-serviceable products. It has also meant externalizing costs wherever possible. Here the terms efficiency and effectiveness are used as the great management thinker Peter Drucker used them.[59] Efficiency is "doing things right," i.e., in this case, accepting the current system and attempting to produce as many products as possible without wasting resources. Efficiency is a noble enough goal, except that it fails to take either dignity or limits to growth into consideration.[60] Effectiveness, in Drucker's sense, means "doing the right thing," which in this case means taking ecological limits to growth, climate change, and sustainability issues seriously—and designing businesses and the economic system as a whole to reflect those realities.

One major hurdle would need to be dealt with in shifting to this more shamanic perspective on business and economics: the financial system. Today's economics, economies, businesses, and, in particular, financial system are focused explicitly and narrowly on growth as measured in financial and profitability terms, and growth in the value of companies' share prices. Far from providing financial support to businesses that need funding, as they originally were, today's financial institutions focus on profiteering, largely for their own gain. Since the 1980s, financialization of the economy has come to take an ever greater share of the economy, despite the fact that much of what many financial institutions do lacks productive value.[61] Too much of finance serves mainly as a speculative market in which the rich get richer, while the poorer (or to use the Occupy movement's term for the majority, "the 99%") get poorer, relatively speaking. To shift to a shamanic mindset on business, then, requires that financial institutions be restored to their original purposes of *supporting* business endeavors with funding, rather than creating manufactured "wealth" without regard to the consequences. Simultaneously, new measures of success, including company and economy effectiveness, would need to be devised that take wellbeing, sustainability, and dignity into account. Many people have been working on such measures, which include the OECD Better Life Index, discussed above, and the Genuine Progress Indicator (GPI).[62] GPI takes both resource enhancement and depletion into account, which our traditional measure of Gross Domestic Product (GDP) does not, and also adds in services that typically do not count in GNP, and detracts problematic activities even when they 'add' economic value.

Wisdom and the Shaman

None of this shift, of course, will be easy, which is why we increasingly need wise leaders—elders of our community (no matter what their age)—to guide this transition from a shamanic perspective. All of these shifts would be difficult, of course, but they would put humanity into a significantly more healthful, dignity-based, and sustainable relationship with nature. Traditional shamans, in addition to being healers, are often considered to be the wise elders in a given community. So it is today with the need for shamans, whether intellectual or acting in other capacities. The world desperately cries out for wise elders—shamans acting as wise business, political, and civil society leaders (not to mention artists, teachers, therapists and healers, of all sorts)—to guide us toward a new vision of how we as human beings can act in the world and with respect to nature and its many aspects.

Linked to the idea of wisdom, defined as the integration of moral imagination (the good), systems understanding (the true), and aesthetic sensibility (the beautiful) in the service of a better world, a shamanic perspective can, I believe be a way for academics, business people, and other leaders to make a positive difference in the world.[63] By bringing a more shamanic perspective to the roles of business and economics in the world, by focusing on wellbeing and collective value for all, and by recognizing the crucial role of dignity in shaping a better world, we can all help to articulate a new set of theories, insights, ideas, and, importantly, memes that shape a more ecologically-grounded worldview and consequently more ecologically grounded economics and business models. Such a perspective would incorporate notions of wellbeing for all, where "all" in the shamanic sense, truly does mean all humans, all other living beings, all aspects of nature, and all of the Earth itself, considered as the living entity Gaia.

Discussion Questions

Explain the concept of intellectual shaman and how it might apply to individuals attempting to achieve sustainability in the world.

How does the idea of intellectual shamanism relate to the definition of wisdom given in the chapter? Do you agree or not with this definition of wisdom? If so, how can you apply this idea in your own life? If not, what definition would you offer and how could you apply that definition in your own life?

How are concepts of wellbeing and dignity for all related to Indigenous beliefs about the world around us? Why might it be important to think about the world in terms of wellbeing and dignity rather than simply (financial) wealth maximization? Who might benefit from such a perspective and how?

Explain the idea of collective value and think about how some business or other enterprise that you are aware of might use such a concept.

Notes

1. Serge Kahili King, *Urban Shaman* (New York, NY: Simon and Schuster, 2009).
2. Igor Gorkov, personal communication, 2016.
3. Sandra Waddock, *Intellectual Shamans: Management Academics Making a Difference* (Cambridge, UK: Cambridge University Press, 2015); Peter J. Frost and Carolyn P. Egri, "The Shamanic Perspective on Organizational Change and Development," *Journal of Organizational Change Management* 7, no. 1 (1994): 7–23; Carolyn Egri and Peter Frost, "Shamanism and Change: Bringing Back the Magic in Organizational Transformation," *Research in Organizational Change and Development* 5 (1991): 175–221; James Dow, "Universal Aspects of Symbolic Healing: A Theoretical Synthesis," *American Anthropologist* 88, no. 1 (1986): 56–69; Sandra Waddock, "Reflections: Intellectual Shamans, Sensemaking, and Memes in Large System Change," *Journal of Change Management* 15, no. 4 (2015): 1–15.
4. Waddock, *Intellectual Shamans*.
5. Mircea Eliade, *Shamanism: Archaic Techniques of Ecstasy* (New York, NY: Pantheon Books, 1964).
6. King, *Urban Shaman*.
7. Waddock, *Intellectual Shamans*; Frost and Egri, "The Shamanic Perspective "; Egri and Frost, "Shamanism and Change."
8. Dow, "Universal Aspects of Symbolic Healing."
9. Frost and Egri, "The Shamanic Perspective."
10. Dow, "Universal Aspects of Symbolic Healing."
11. Frost and Egri, "The Shamanic Perspective"; Egri and Frost, "Shamanism and Change."
12. Richard Dawkins, *The Selfish Gene*, 30th Anniversary ed. (New York, NY: Oxford University Press, 2006). See Sandra Waddock, "Wisdom and Responsible Leadership: Aesthetic

Sensibility, Moral Imagination, and Systems Thinking," in *Aesthetics and Business Ethics: Issues in Business Ethics*, ed. Daryl Koehn and Dawn Elm (New York, NY: Springer, 2015).

13. Susan J. Blackmore, *The Meme Machine* (New York, NY: Oxford University Press, 1999).

14. Intergovernmental Panel on Climate Change, *Climate change 2013: The Physical Science Basis. Working Group I Contribution to the 5th Assessment Report of the Intergovernmental Panel on Climate Change: Changes to the Underlying Scientific/Technical Assessment to Ensure Consistency with the Approved Summary for Policymakers* (Geneva, Switzerland: United Nations, 2013); Paul Gilding, *The Great Disruption: How the Climate Crisis Will Transform the Global Economy* (London, UK: Bloomsbury, 2011); James Lovelock, *The Vanishing Face of Gaia: A Final Warning*, pbk. ed. (New York, NY: Basic Books, 2010).

15. Thomas Piketty, *Capital in the 21st Century* (Cambridge, MA: Harvard University Press, 2014).

16. Rakesh Kochhar, "A Global Middle Class Is More Promise Than Reality: From 2001 to 2011, Nearly 700 Million Step Out of Poverty, but Most Only Barely," (Washington, D.C.: Pew Research Center, 8 July 2015), updated 13 August 2015, accessed 16 August 2018, http://www.pewglobal.org/2015/07/08/a-global-middle-class-is-more-promise-than-reality/.

17. Pew Research Center, "The American Middle Class Is Losing Ground: No Longer the Majority and Falling Behind Financially," (Washington, D.C.: Pew Research Center, 9 December 2015), accessed 16 August 2018, http://www.pewsocialtrends.org/2015/12/09/the-american-middle-class-is-losing-ground/.

18. Jared M. Diamond, *Collapse: How Societies Choose to Fail or Succeed* (New York: Viking, 2005).

19. Dow, "Universal Aspects of Symbolic Healing;" Four Arrows, *Point of Departure: Returning to Our More Authentic Worldview for Education and Survival* (Charlotte, NC: Information Age, 2016).

20. James E. Lovelock, *Gaia, a New Look at Life on Earth* (New York, NY: Oxford University Press, 1979).

21. Fritjof Capra, "Complexity and Life," *Theory, Culture & Society* 22, no. 5 (2005): 33–44; *The Web of Life: A New Scientific Understanding of Living Systems*, 1st Anchor Books ed. (New York: Anchor Books, 1996); *The Systems View of Life: A Unifying Vision*, ed. Pier Luigi Luisi (Cambridge, UK: Cambridge University Press, 2014).

22. Donna Hicks, *Dignity: The Essential Role It Plays in Resolving Conflict* (New Haven, CT: Yale University Press, 2011).

23. Ibid.

24. Ibid.

25. Amartya Sen, "Well-Being, Agency, and Freedom: The Dewey Lectures 1984," *The Journal of Philosophy* 82, no. 4 (1985): 169–221.

26. Capra, *The Web of Life*; "Complexity and Life;" *The Systems View of Life*.

27. Dow, "Universal Aspects of Symbolic Healing;" Arrows, *Point of Departure*.

28. Lovelock, *Gaia, a New Look at Life on Earth*; *The Vanishing Face of Gaia*.

29. Hicks, *Dignity*.

30. Gro Harlem Brundtland, "Our Common Future: World Commission on Environmental Development," in *The Brundtland Report* (Oxford, UK: United Nations, 1987).

31. Molly Larkin, "What Is the 7th Generation Principle and Why Do You Need to Know About It?" *Ancient Wisdom for Balanced Living* [blog], (15 May 2013).

32. Sandra Waddock, "We Are All Stakeholders of Gaia: A Normative Perspective on Stakeholder Thinking," *Organization & Environment* 24, no. 2 (2011): 192–212.

33. Thomas Donaldson and James P. Walsh, "Toward a Theory of Business," *Research in Organizational Behavior* 35 (2015): 181–207.

34. Donaldson & Walsh, "Toward a Theory of Business," 188.

35. Ibid., 188.

36. Ibid., 192.

37. Ibid., 196.

38. Ibid., 197.

39. Ibid., 198.

40. E.g. Michael Pollan, *The Omnivore's Dilemma: A Natural History of Four Meals* (New York, NY: Penguin, 2006).

41. Eric Schlosser, *Fast Food Nation: The Dark Side of the All-American Meal* (Boston, MA: Houghton Mifflin, 2001); Pollan, *The Omnivore's Dilemma*.

42. Pollan, *The Omnivore's Dilemma*; Schlosser, *Fast Food Nation*.

43. Schlosser, *Fast Food Nation*.

44. Pollan, *The Omnivore's Dilemma*.

45. E.g. Michael Moss, *Salt, Sugar, Fat: How the Food Giants Hooked Us*, 1st ed. (New York, NY: Random House, 2013); Pollan, *The Omnivore's Dilemma*.

46. Peter F. Drucker, *The Effective Executive*, 1st ed. (New York, NY: Harper & Row, 1967).

47. Moss, *Salt, Sugar, Fat*.

48. Sébastien Sauvé, Sophie Bernard, and Pamela Sloan, "Environmental Sciences, Sustainable Development and Circular Economy: Alternative Concepts for Trans- Disciplinary Research," *Environmental Development* 17 (2016): 48–56.

49. Leading for Wellbeing Consortium. Leading for Wellbeing [website]. Accessed 16 August 2018. http://leading4wellbeing.org/.

50. Romina Boarini and Marco Mira Ercole, "Going Beyond GDP: An OECD Perspective," *Fiscal Studies* 34, no. 3 (2013), 293.

51. Boarini & d'Ercole, "Going Beyond GDP," 292.

52. Ibid., 294–295; Martine Durand, "The OECD Better Life Initiative: How's Life? And the Measurement of Well-Being," *Review of Income and Wealth* 61, no. 1 (2015): 7–8.

53. Durand, "The OECD Better Life Initiative."

54. Sandra Waddock, "Generative Businesses Fostering Vitality: Rethinking Business' Relationship to the World," (Working Alternatives, 2016).

55. Pollan, *The Omnivore's Dilemma*.

56. Janine M. Benyus, *Biomimicry: Innovation Inspired by Nature*, 1st ed. (New York, NY: Morrow, 1997).

57. Thomas Homer-Dixon, "*The Ingenuity Gap: Facing the Economic, Environmental, and Other Challenges of an Increasingly Complex and Unpredictable World*," (New York, NY: Intergovernmental Panel on Climate Change, 2002); Gilding, *The Great Disruption*.

58. John Ehrenfeld and Andrew Hoffman, *Flourishing: A Frank Conversation about Sustainability* (Stanford, CA: Stanford University Press, 2013); Tim Jackson, *Prosperity without growth: Economics for a finite planet* (Abington, UK: Routledge, 2011).

59. Drucker, *The Effective Executive*.

60. Donella H. Meadows, William W. Behrens III, and Jørgen Randers, *The Limits to Growth: A Report for the Club of Rome's Project on the Predicament of Mankind* (New York, NY: Universe Books, 1972); Graham M. Turner, "A Comparison of the Limits to Growth with 30 Years of Reality," *Global Environmental Change* 18, no. 3 (2008): 397–411; Jorgan Randers, *2052: A Global Forecast for the Next Forty Years* (White River Junction, VT: Chelsea Green, 2012).

61. Benjamin Landy, "Graph: How the Financial Sector Consumed America's Economic Growth," *The Century Foundation*, february 25, 2013, accessed onlie, 22 August 2018, https://tcf.org/content/commentary/graph-how-the-financial-sector-consumed-americas-economic-growth/?agreed=1.

62. Redefining Progress [website], Accessed online, 26 August 2015, https://community-wealth.org/content/redefining-progress. http://rprogress.org/sustainability_indicators/genuine_progress_indicator.htm.

63. Waddock, "Wisdom and Responsible Leadership."

References

Benyus, Janine M. *Biomimicry: Innovation Inspired by Nature* (1st ed.). New York, NY: Morrow, 1997.

Blackmore, Susan J. *The Meme Machine*. New York, NY: Oxford University Press, 1999.

Boarini, Romina, and Marco Mira Ercole. "Going Beyond GDP: An OECD Perspective*." *Fiscal Studies* 34, no. 3 (2013): 289–314.

Brundtland, Gro Harlem. "Our Common Future: World Commission on Environmental Development." In *The Brundtland Report*. Oxford, UK: United Nations, 1987.

Capra, Fritjof. "Complexity and Life." *Theory, Culture & Society* 22, no. 5 (2005): 33–44.

———. *The Systems View of Life: A Unifying Vision*. Edited by Pier Luigi Luisi. Cambridge, UK: Cambridge University Press, 2014.

———. *The Web of Life: A New Scientific Understanding of Living Systems*. 1st Anchor Books ed. New York, NY: Anchor Books, 1996.

Dawkins, Richard. *The Selfish Gene*. 30th Anniversary ed. New York, NY: Oxford University Press, 2006.

Diamond, Jared M. *Collapse: How Societies Choose to Fail or Succeed*. New York, NY: Viking, 2005.

Donaldson, Thomas, and James P. Walsh. "Toward a Theory of Business." *Research in Organizational Behavior* 35 (2015): 181–207.

Dow, James. "Universal Aspects of Symbolic Healing: A Theoretical Synthesis." *American Anthropologist* 88, no. 1 (1986): 56–69.

Drucker, Peter F. *The Effective Executive* (1st ed.). New York, NY: Harper & Row, 1967.

Durand, Martine. "The OECD Better Life Initiative: How's Life? And the Measurement of Well-Being." *Review of Income and Wealth* 61, no. 1 (2015): 4–17.

Egri, Carolyn, and Peter Frost. "Shamanism and Change: Bringing Back the Magic in Organizational Transformation." *Research in Organizational Change and Development* 5 (1991): 175–221.

Ehrenfeld, John, and Andrew Hoffman. *Flourishing: A Frank Conversation about Sustainability*. Stanford, CA: Stanford University Press, 2013.

Eliade, Mircea. *Shamanism: Archaic Techniques of Ecstasy*. New York, NY: Pantheon Books, 1964.

Four Arrows. *Point of Departure: Returning to Our More Authentic Worldview for Education and Survival*. Charlotte, NC: Information Age, 2016.

Frost, Peter J., and Carolyn P. Egri. "The Shamanic Perspective on Organizational Change and Development." *Journal of Organizational Change Management* 7, no. 1 (1994): 7–23.

Gilding, Paul. *The Great Disruption: How the Climate Crisis Will Transform the Global Economy*. London, UK: Bloomsbury, 2011.

Hicks, Donna. *Dignity: The Essential Role It Plays in Resolving Conflict*. New Haven, CT: Yale University Press, 2011.

Homer-Dixon, Thomas. *"The Ingenuity Gap: Facing the Economic, Environmental, and Other Challenges of an Increasingly Complex and Unpredictable World."* New York, NY: Intergovernmental Panel on Climate Change, 2002.

Intergovernmental Panel on Climate Change. *Climate Change 2013: The Physical Science Basis. Working Group I Contribution to the 5th Assessment Report of the Intergovernmental Panel on Climate Change: Changes to the Underlying Scientific/Technical Assessment to Ensure Consistency with the Approved Summary for Policymakers*. IPCC-36 / Doc.4, Agenda Item 3. Geneva, Switzerland: United Nations, 27 September 2013.

Jackson, Tim. *Prosperity without Growth: Economics for a Finite Planet*. Abington, UK: Routledge, 2011.

King, Serge Kahili. *Urban Shaman*. New York, NY: Simon and Schuster, 2009.

Kochhar, Rakesh. "A Global Middle Class Is More Promise Than Reality: From 2001 to 2011, Nearly 700 Million Step Out of Poverty, but Most Only Barely." Washington, D.C.: Pew Research Center, 8 July 2015. Updated 13 August 2015. Accessed 16 August 2018. http://www.pewglobal.org/2015/07/08/a-global-middle-class-is-more-promise-than-reality/

Landy, Benjamin. "Graph: How the Financial Sector Consumed America's Economic Growth ." *The Century Foundation*. 25 February, 2013. https://tcf.org/content/commentary/graph-how-the-financial-sector-consumed-americas-economic-growth/?agreed=1.

Larkin, Molly. "What Is the 7th Generation Principle and Why Do You Need to Know About It?" Ancient Wisdom for Balanced Living [blog]. 15 May 2013. Accessed 16 August 2018. https://www.mollylarkin.com/what-is-the-7th-generation-principle-and-why-do-you-need-to-know-about-it-3/.

Leading for Wellbeing Consortium. Leading for Wellbeing Coalition [website]. Accessed 16 August 2018. http://leading4wellbeing.org/.

Lovelock, James E. *Gaia, a New Look at Life on Earth*. New York, NY: Oxford University Press, 1979.

———. *The Vanishing Face of Gaia: A Final Warning*. Pbk. ed. New York, NY: Basic Books, 2010.

Meadows, Donella H., William W. Behrens III, and Jørgen Randers. *The Limits to Growth: A Report for the Club of Rome's Project on the Predicament of Mankind*. New York, NY: Universe Books, 1972.

Moss, Michael. *Salt, Sugar, Fat: How the Food Giants Hooked Us* (1st ed.). New York, NY: Random House, 2013.

Pew Research Center. "The American Middle Class Is Losing Ground: No Longer the Majority and Falling Behind Financially." Washington, D.C. 9 December 2015. Accessed 16 August 2018. http://www.pewsocialtrends.org/2015/12/09/the-american-middle-class-is-losing-ground/.

Piketty, Thomas. *Capital in the 21st Century*. Cambridge, MA: Harvard University Press, 2014.

Pollan, Michael. *The Omnivore's Dilemma: A Natural History of Four Meals*. New York, NY: Penguin, 2006.

Randers, Jorgan. *2052: A Global Forecast for the Next Forty Years. White River Junction*. White River Junction, VT: Chelsea Green, 2012.

Redefining Progress [website]. Genuine Progress Indicator. Accessed online 26 August 2018. http://rprogress.org/sustainability_indicators/genuine_progress_indicator.htm.

Sauvé, Sébastien, Sophie Bernard, and Pamela Sloan. "Environmental Sciences, Sustainable Development and Circular Economy: Alternative Concepts for Trans-Disciplinary Research." *Environmental Development* 17 (2016): 48–56.

Schlosser, Eric. *Fast Food Nation: The Dark Side of the All-American Meal*. Boston, MA: Houghton Mifflin, 2001.

Sen, Amartya. "Well-Being, Agency, and Freedom: The Dewey Lectures 1984." *The Journal of Philosophy* 82, no. 4 (1985): 169–221.

Turner, Graham M. "A Comparison of the Limits to Growth with 30 Years of Reality." *Global Environmental Change* 18, no. 3 (2008): 397–411.

Waddock, Sandra. "Generative Businesses Fostering Vitality: Rethinking Business' Relationship to the World." Boston College Working Paper, 2016.

———. *Intellectual Shamans: Management Academics Making a Difference*. Cambridge, UK: Cambridge University Press, 2015.

———. "Reflections: Intellectual Shamans, Sensemaking, and Memes in Large System Change." *Journal of Change Management* 15, no. 4 (2015): 1–15.

———. "We Are All Stakeholders of Gaia: A Normative Perspective on Stakeholder Thinking." *Organization & Environment* 24, no. 2 (2011): 192–212.

———. "Wisdom and Responsible Leadership: Aesthetic Sensibility, Moral Imagination, and Systems Thinking." In *Aesthetics and Business Ethics: Issues in Business Ethics*, edited by Daryl Koehn and Dawn Elm, 129–47. New York, NY: Springer, 2015.

CHAPTER THIRTEEN

For a Tattered Planet

Art and Tribal Continuance

KIMBERLY BLAESER

Art opens us to the sacred, a periodic reentry into time primordial.

—Jane Ash Poitras[1]

Art and Endurance

When icebergs are breaking off in the Antarctica and global populations continue to clamor for more oil, why art? When throughout the world Indigenous peoples still suffer at the hands of colonial powers, why in their struggle for justice do members of these nations turn to poetry and music? Audre Lorde once claimed, "Poetry is not a luxury."[2] How then might the arts be considered a necessity? In what ways do they play a role in preserving, voicing, and disseminating the "sustainable wisdom" of Native nations? I believe the arts can touch and teach, can heal, inspire, and bring change.

Paul Chaat Smith, Comanche author and art curator, says, "Artists are deeply respected in the Native World. We ask of them just two things: (1) make fabulous art, and (2) lead the revolution."[3] Chaat Smith's statement is playful yet heartfelt. Curator and Native art historian Bruce Bernstein states with more gravitas, "Native art plays a central role in health and endurance ... Art is why tribes persist, as it strengthens Native American aspects of life."[4] Both Chaat Smith's and

Bernstein's comments align with the declarations of Native artists and intellectuals from Jaune Quick-To-See-Smith to Edgar Heap of Birds, from Victor Masayesva to Colleen Cutschall, many of whom have pointed to the key role art plays in the survival and flourishing of Native cultures.

Jane Ash Poitras, for example, claims, "One can be subjected to many inhumane conditions, but, for the survival of the soul, the mind is always free through art." Heap of Birds states: "For today's Native artist, it is imperative to pronounce strong personal observations concerning the individual and political conditions that we experience."[5] But how, specifically, do we understand the work we do as Native writers, artists, musicians, photographers, weavers, storytellers, etc.? How does the creative work function within our communities? In particular, how does or how might that creative process sustain wisdom, sustain culture, and sustain life?

Some days my eighty-six billion neurons come alive processing theories about aesthetics, epistemology, and creativity. Other days and dream-filled nights, I give myself over to intuition, to muse, to spirit—to the hunger to enter the artistic process. When recently asked to respond in three to five sentences to the prompt, "I write in order to ...," I evaded the strict direction of the ellipses. Confusion, inspiration, and compulsion have fueled my work in equal parts. I began writing in order to find a survivable pattern in the jagged puzzle that was my life in a struggling Native community. That search itself became the path, the pattern. Writing is ultimately an act of attention, a process of seeing and then re-seeing. So, I write because the world reveals itself, but the world reveals itself partly through the search in language for meaning, understanding, and beauty. (You knew if we probed long enough we would reach the spiral of being. We always do.) Nothing in itself. Everything a process. Every process a reciprocity.

Like many Native people, I engage in the arts as a way of being in the world—a path of both search and service. Today that often means, as Chaat Smith suggests, that art is spectacle but also fuels tribal resistance to conditions such as injustice or ecological degradation, it fuels our work for change. According to Bernstein, art "holds the message of the past, and proposes and situates the continuity of the individual and Native communities for the future."[6] It becomes a path for continuance.

I often recall lines from Chickasaw writer Linda Hogan's poem "Neighbors" as I contemplate the balance I and other Native writers or artists try to maintain. She writes, "This is the truth, not just a poem."[7] Our challenge: make truth with our art. But, she also writes, "This is a poem and not just the truth."[8] The inherent challenge in this statement is nothing less than artistic beauty itself. Together these lines suggest the aesthetic foundation of my own work (these goals are met to a greater or lesser degree in individual pieces), one that includes both the affective and effective reality of art. Through "creative acts," I strive for these linked aspirations: to make my work beautiful as language, as art; and at the same time, to engage through the art in the sustainable workings of the world.

There are many ways the art of poetry "makes return" in the world. For example, it can: heighten our awareness of the everyday, gesture towards the unspeakable, offer us solace, provide entertainment, play with language, re-member the past, encapsulate the beautiful, pull back the mask from the dark side of existence, celebrate love, critique, work for change, sing, chant, taunt, and even lead us to the other side of language—to silence or to a more soulful, vibrant encounter with the world.

The pathway of poetic revelation is succinctly described by Lorde:

> It forms the quality of the light within which we predicate our hopes and dreams toward survival and change, first made into language, then into idea, then into more tangible action. Poetry is the way we help give name to the nameless so it can be thought.[9]

Of course, often the creative acts feed us in several ways simultaneously. Of these many undertakings of art, I focus in the pieces below on two in particular—on acts of spiritual connection (often symbolized in the layers of our physical world and its creatures) and on acts of survival (which frequently involve resistance). Each of these—spiritual connection and survival—is a crucial element in what we call "sustainable wisdom."

Pathways to Transcendence

Indigenous peoples have long employed myth and metaphor in tribal teachings, storying the mysterious workings of the universe to create understanding, and metaphorically linking the physical realities with the spiritual—or revealing their always already interconnected reality—through symbolism. On a good day, I am able to connect the creative work I do to that multivalent continuum. I strive to invite not only a seeing or a hearing of what is on the page or in a performance, but to invite a self-perpetuating re-seeing and re-hearing that can lead to the not stated or not present. In the way I understand and attempt to create with them, poetry and photography are both about question and gesture. I see them as allusive—an entrance into a process of engagement with the images and ideas, into a discovery of meaning(s), perhaps into a path toward enlightenment.

Among my artistic "gestures," I sometimes include elements of tribal story as one layer within a poem or picto-poem. The piece below, for example, involves an ancient myth central to Native understanding of our reciprocal relationship to this earth we call Turtle Island—that of the Earthdiver. Common to peoples all over the world, the key elements of the story include a flood and the need to gather grains of sand from which to build a new earth. Tellings vary, but the Anishinaabeg teachings emphasize the significance of all—small or large beings—and how we have to work together in the continuing creation and recreation of the world—a powerful responsibility. (In this they distinguish themselves from Biblical creation stories in which the world is created *for* human beings, in which the world is ordered hierarchically). The Anishinaabeg story features in particular the self-sacrificing action of one of the smallest of water beings—*Wazhashk* or muskrat. But the poem wants in its allusiveness to also emphasize our human insignificance and, at the same time, our relatedness, our belonging to something more vast than ourselves—to the immense bodies of water, to story, to all-time. I use the surreal rendition of "water beings" in the photo art to untether reader/viewer from "reality" or so-called "realism," to invite a re-imagining of their own "dream" self, perhaps a realignment of their own "mythic quest."

As demonstrated in "Dreams of Water Bodies" (see Figure 13.1), my own experience of the natural world arises partly from and is imbued with a tribal context involving ancestral connections, tribal continuance, and spiritual continuities; involving a cultural framework of origin stories, ceremony, and ritual. In a "true" or successful rendering of natural places then, in my work they must come alive as *Native spaces*. Just as Lakota artist Colleen Cutschall speaks of "drawing where the seen and unseen come together to form the narrative," and Diné storyteller Yellowman speaks about the role of Native creative works noting how they "allow us to envision the possibility of things not ordinarily seen or experienced" and to use "the intangible part of our thinking mind," I, too, endeavor through image, presentation, and content to insinuate the animate interconnected reality of tribal experience.[10]

Specifically in terms of form, I work towards implying the multi-dimensionality by literally employing several senses, by joining—sometimes juxtaposing or layering—text and image. I use symbolic visual cues: reflection and refraction of image, shadow, suggestion of movement; employ visual disruption including an advertised absence or non-representational images; and employ concrete poetry, juxtapositions of language, and interactive textual allusions.

Although this listing of techniques suggests a plodding cerebral process, I am only able to construct this analysis looking back at an assemblage of characteristic work. In fact, I actually create toward a sensibility—by hearkening after a certain slant of knowing. The image of the next picto-poem (See Figure 13.2), for example, began with the Great Blue Heron flying through blue sky. The photo was made during a paddle with my son in the Horicon Marsh in Wisconsin. On Mother's Day in 2012, we spent many golden minutes watching the heron stalk fish, catch and swallow them whole. We watched it fly—with great pomp in lift-offs and landings, with sometimes a yellow-bellied *giigoonh* still carried in its beak.

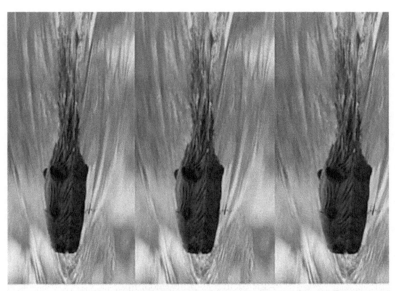

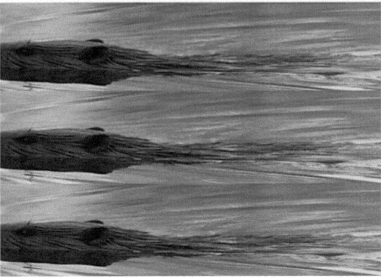

Dreams of Water Bodies Nibii-wiiyawan Bawaadanan*

Wazhashk,	Wazhashk
small whiskered swimmer,	agaashiinyi memiishanowe bagazod
you, a fluid arrow crossing waterways	biwak-dakamaadagaayin
with the simple determination	mashkawndamyin
of one who has dived	googiigwaashkwaniyamban
purple deep into mythic quest.	dimii-miinaandeg gagwedweyamban.
Belittled or despised	G'goopazomigoog
as water rat on land;	ninii-chiwaawaabiganoojinh akiing
hero of our Anishinaabeg people	ogichidaa Anishinaabe
in animal tales, creation stories	awesiinaajimowinong, aadizookaang
whose tellers open slowly,	dash dibaajimojig onisaakonanaanaawaa
magically like within a dream,	nengatch enji-mamaanjiding
your tiny clenched fist	gdo'bikwaakoninjens
so all water tribes.	midash kina Nibiishinaabe
might believe.	debwe'endamowaad.
See the small grains of sand—	Waabandanan negawan
Ah, only those poor few—	aah sa ongow eta
but they become our turtle island	maaaji-mishiikenh-minis
this good and well-dreamed land	minwaabandaan aakiing maampii
where we stand in this moment	niigaanigaabawiying
on the edge of so many bodies of water	agamigong
and watch *Wazhask,* our brother,	Wazhashk waabamang, niikaanaanig
slip through pools and streams and lakes	zhiibaasige zaaga'iganan gaye ziibiinsan
this marshland earth hallowed by	mashkiig zhawedaamin
the memory	mikwendamin
the telling	waawiindamin
the hope	ezhi-bagosenimowaad
the dive	ezhi-googiiwaad
of sleek-whiskered-swimmers	agaashiinyag memiishanowewaad bagazojig
who mark a dark path.	dibiki-miikanong.
And sometimes in our water dreams	Nangodinong enji-nibii-bawaajiganan
we pitiful land-dwellers	gidimagozijig aakiing endaaying
in longing	bakadenodamin
recall, and singing	dash nagamoying
make spirits ready	jiibenaakeying
to follow:	noosone'igeying
*bakobii.***	bakobiiying.

**Go down into the water.   *Translation by Margaret Noodin

Figure 13.1: Picto-Poem "Dreams of Water Bodies."
Source: Author.

The feel of the encounter was rapturous; however, the image, when I first printed it, was an anemic half rendering. I didn't know why. The position of the bird was stunning—huge wingspan and wingtips curving up, neck in a classic curve, legs extended, and feet pointing straight out to forever. When I thought about what I saw and what the camera "captured," I realized we don't ever only see what we see. What is literally before us is always experienced in the context of our own understandings and beliefs. I realized that heron had flown through the territories of Anishinaabeg ontology and mythology—as *ajijaak* the echo-maker and totemic figure for the crane clan. How could I make the image suggest the context of my layered reality? I employed gesture, made symbolic use of Anishinaabe floral beadwork together with snatches of language including the bird's name in Anishinaabemowin—*Zhashagi*. Ultimately, the lines of the poem suggests the search I made for a way to give voice to my encounter, to the "knowledge" that Native spaces still exist in our everyday landscapes, to my understanding of how readily we can re-inhabit them.

Ancient light abides "where the seen and unseen come together," in "the intangible part of our thinking mind."[11] Ultimately, the picto-poem—with *Zhashagi* flying through a sky of woodland beadwork—disrupts expectations. "Reading" the image or the poem requires the reader/viewer to inhabit "an other" perspective. It attempts to create a pathway toward this alternate way of knowing, toward a moment of transcendence.

Embodying Poetic Resistance

Transcendence in poetry, although philosophically uncomfortable for some reader/listeners, in some ways requires less accommodation than the challenges brought by Native resistance poetry. The "truth" Hogan speaks of and the "tangible action" Lorde suggests, first run afoul of the "art for art's sake" aesthetic espoused by some poets and scholars, and also tend in this context—by their origin in "an other" (here Indigenous) worldview—to contest generally accepted or mainstream value systems and cultural life ways. Specifically in the arena of "sustainable wisdom," for example, communal values may be set against capitalistic ideals. Likewise, Native notions of reciprocity, kinship, and the animate earth do not align with resource "ownership" or the labels of "inanimate" and "thing" often used when categorizing elements of the natural world. Following from these differences are any number of "tangible actions" from protecting the water resources at Standing Rock to working to prevent development of sacred sites such as effigy mounds.

A Crane Language

From where behind you ancient light
enters. Curves of this woodland sky,
a fissure in the propriety of reason.
Stitched here is first translucence. Sun
turning the hose of your throat to a vessel
of fire. Body angled like a sundial
against the cup of all reluctance. Our rock cold
clan longing. This silence. Released now
from the weighted anchor of time.
Your voice—a torch of language. *Zhashagi.*

The flame of your tongue gives light.

Figure 13.2: Picto-Poem "A Crane Language."
Source: Author.

The impetus for the particular "ecocritical" poem I include below (Figure 13.3) does indeed involve both "truth" and "tangible action." It focuses on the impact of environmental actions and the "truth" of the Native philosophy and practice of making decisions while considering the long view—seven generations in the past and seven generations into the future. The "tangible action" the poem calls for is standing: standing up for those impacted by environmental racism, standing against the companies and legislators who threaten our land and water, standing with those who oppose environmental degradation. The piece might also be categorized as "documentary" poetry, making use, as it does, of historical and political details and newspaper accounts surrounding the debate of the proposed open-pit iron mine in the Penokee Hills of Wisconsin. Of course, the poem also contains other allusions playing, for example, on a mathematical equation with multivariable functions and linking the current environmental case to others in history.

The kind of historic contextualizing of Indigenous resistance I undertake in the poem matters in nations like the United States where colonial forgetfulness has become a middle-class pastime. In her essay "Places to Stand: Art after the American Indian Movement," scholar Jessica Horton discusses this need to historicize contemporary Indigenous struggles through art saying, "For artists who witnessed the forgetting of AIM, the vitality of contemporary struggles rests on uncovering historical ground."[12] Indeed, Horton suggests the appropriate historical context is "a vast colonial story implicating Indigenous bodies and territory for more than five centuries."[13]

Although my task in this discussion is not to characterize in detail Native cosmologies, belief systems, or foundational community practices related to sustainability and spirituality; as I have noted, certain key epistemological concepts do inform my work. Among those of particular significance to the pieces included: concepts of motion, repetition, or cycle; an understanding of overlapping or shared realities; totemic animals; the significance of verbal restraint and absence; autonomous reciprocal relationships; and an emphasis on community.

Indeed, community weighs especially greatly in the origin of this essay and in the title poem shared herein. Joe Horse Capture once wrote about the importance of "communal sharing and exchange" claiming, "To learn together is to keep the People together."[14] As writer-in-residence for the "Sustainable Wisdom" conference at Notre Dame, one of my tasks was to create a poem for the conference. In order to do that in a way that reflected the community learning of those several days in September 2016, I elected to create a poem *from* the conference, one that wove together ideas and even bits of language spoken during presentations and

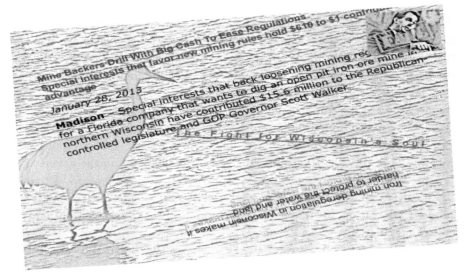

Of the many ways to say: *Please Stand*

I. ∂,Partial Differential Equation

What we erase from polite conversation. Bodies on fire. The historic cleansing of the landscape, the sweep of humanity west, west, west. Environmental r ism.

All things being equal, things are never equal. Think of scope. Like the reach of the imperial. Or consider variables. Value. Or commodity. Convenience policy, a tally mark across generations. Uranium mining. Atomic bomb detonation at White Sands. A complicated table of fallout factors. Plume of greatness.

Shift. Angles and perspectives. Ways of seeing. *Seven generations into the past; seven generations into the future.*

Or how to solve for survival.

II. *Zongide'en*, Be Brave.

Another partial differential equation. Let's say a corporation proposes a mine. Variables include Tyler Forks. Bad. Potato. Rivers. A 22-mile, 22,000-acre strip of land. Jobs. *Maanomin*. Open pit. Exceptional or outstanding resource waters. Legislation. Iron oxide. Fish. Blasting and pulverizing. New legislation.

The functions depend upon the continuous variables. Fluid flow, for example. And changing laws. Somewhere along the granite line, someone enters. Let's say they have put down one life and taken up another: the solution of the PDE.

Warriors (walleye, Indian, new-age) face arbitrary functions. Changing laws. Guards. Guns. If life is stretched over two points. It vibrates. We cannot measure that vibration in this generation.

We can sing it, or make it into light.

Figure 13.3: Picto-Poem "Of the many ways to say: *Please Stand.*"
Source: Author.

discussions. Although there is a longstanding tradition in poetry for "found" poems created from language culled from other sources, this poem came into being accumulating ideas and sometimes imagery from speakers as well. For their permission to build in this way, I extend my gratitude to fellow conference participants.

This poem, like the other three included here, invites a re-vision of attitudes toward our planet and offers both a new narrative of our environmental history and an alternate path for our future interactions with our earth. It suggests a path toward continuance, one that spirals among tenets of Indigenous wisdom on sustainability. In our dealings with this Turtle Island of mythical, spiritual, and physical being, it asks of us a different measure.

Poem for a Tattered Planet: If the Measure is Life

Born
under the canopy of plenty
    the sweet unfolding
    season's of a planet's youth,
in the trance of capitalism we take our fill
content with the status quo
pull our shades on encroaching collapse
say something about Anthropocene,
the energy barter and the holy fortress of science.

But beyond
the throat of commerce,
beneath the reflection
        of the celestial river,
within the ancient copper beauty of belonging
we stand encircled
    inhabit the Ish,
navigate by the singing of songs.

Though money fog settles around,
confounds measure
today veil of mystery   shifts
lifts for momentary sigh t.
        Here
find rhythm of a tattered planet,
feel on panther mound

a pulse.   Listen—don't count.
Feel small life drum beneath ____.

My core. I am ancient refracted light
or sound
traveling,
my frequency a constant
my voice
bending at angles
to become whole in another surface—
say a poem.

Say a poem
perpendicular to the boundary
of meaning,
make it a prism or possibility
sing of turtle or cast the mythic lumen
of thunderbird here
on the flat f alter of words:

This page   not contract
but covenant.
Neither image nor voice
will twin itself
In the thick moist   cloud
of being
each limb a nimble test   of tree
glimpse not see   nor calculate.

Still Shroud of Commerce shrouds meaning.
In the technology of documentary genocide
in the destructive bonanza of the industrial age—
declare the death of planet
as it passes   through a sound speed gradient
comes out   on the other side
a lost echo of human greed
repeating itself
repeating itself
repea t in g

Each splinter of language
bent   in complicated formulas of inference

as fog forgets   then remembers   form
and we measure   in metaphor
vibration   earth timbre
refractive errors
of science   or prayer.

Speak the ninety-nine names for god:
Gizhe-manidoo, Great Spirit, or longing,
trembling aspen, the sung bones of salmon,
braided sweetgrass,
the sacred hair bundles of women,
this edible landscape—
aki, nabi, ishkode, noodin,
the ten little winds of our whirled fingertips,
this round dance of the seasons—

the ineffable flourishing.

With mind as holy wind
and voice a frog's bellowous night song
we arrive.
Here sandhill cranes mark sky.
If the measure is life—
their clan legs the length of forever.
Here mirror of lake a canvas of belief.
If the measure is life—
refraction the trigger of all knowing.
Only this.

Now we place *aseema,*
the fragrant tobacco bodies of our relatives
To make the tattered whole.
A question of survival.

Of correlation.

                              Of vision.

The measure is life.

Discussion Questions

What roles have the arts historically played in preserving Indigenous cultures? What roles do contemporary Native artists suggest they play today?

Can works of poetry or the other arts reveal or imply something not stated or not pictured? If so, how might they accomplish this?

How do our cultural perspectives influence the way we see or experience the world on an everyday basis? What mythic underpinnings do we bring to the creation or understanding of our own life story?

What examples can you give of art addressing ecological threats to our planet?

Regardless of your current position or employment—whether you are a student, writer, artist, biologist or computer programmer, how you might describe the relationship between ecological wisdom and sustainability and the work you do?

Notes

1. Jane Ash Poitras, "Paradigms for Hope and Posterity: Wohaw's Sun Dance Drawing," in Plains Indian Drawings, 1865–1935, ed. Jane Catherine Berlo (New York, NY: Harvey N. Abrams, 1996), 68.
2. Audre Lorde, "Poetry is Not a Luxury," in her Sister Outsider: Essays and Speeches (Freedom, CA: Crossing Press, 1984), 36.
3. Paul Chaat Smith, "Famous Long Ago," in Shapeshifting: Transformations in Native American Art, ed. Karen Kramer Russell (New Haven, CT: Yale University Press, 2012), 215.
4. Bruce Bernstein, "Expected Evolution: The Changing Continuum," in Shapeshifting: Transformations in Native American Art, ed. Karen Kramer Russell (New Haven, CT: Yale University Press, 2012), 30.
5. Jane Ash Poitras, "Paradigms for Hope and Posterity"; Edgar Heap of Birds, "Of Circularity and Linearity in the Work of Bear's Heart," in Plains Indian Drawings, 1865–1935, ed. Jane Catherine Berlo (New York, NY: Harvey N. Abrams, 1996), 66, 68.
6. Bernstein, "Expected Evolution," 30.
7. Linda Hogan, "Neighbors," in Savings (Minneapolis, MN: Coffee House Press, 1988), 65. Lines from "Neighbors" are reprinted by permission from Savings (Minneapolis, MN: Coffee House Press, 1988); copyright 1988 by Linda Hogan.
8. Linda Hogan, Savings (Minneapolis, MN: Coffee House Press, 1988), 65.
9. Lorde, "Poetry is Not a Luxury," 35.
10. Colleen Cutschall, "The Seen and the Unseen to Form the Narrative …," in Plains Indian Drawings, 1865–1935, ed. Jane Catherine Berlo (New York, NY: Harvey N. Abrams, 1996), 64–65.
11. Ibid.

12. Jessica L. Horton, "Places to Stand: Art after the American Indian Movement," *Wasafiri: Native North American Literature and Literary Activism* 32, no. 2 (2017), 30.
13. Horton, "Places to Stand," 30.
14. Joe D. Horse Capture, "Time-Honored Expression: The Knowing of Native Objects," in *Plains Indian Drawings, 1865–1935*, ed. Jane Catherine Berlo (New York, NY: Harvey N. Abrams, 1996), 76.

References

Beck, Peggy V., Anna Lee Walters, and Nia Francisco. *The Sacred: Ways of Knowledge, Sources of Life*. Tsaile, AZ: Navajo Community College Press, 1977. Reprint, Flagstaff, AZ: Northland Publishing, 1990.

Bernstein, Bruce. "Expected Evolution: The Changing Continuum." In *Shapeshifting: Transformations in Native American Art*, edited by Karen Kramer Russell, 30–43. New Haven, CT: Yale University Press, 2012.

Chaat Smith, Paul. "Famous Long Ago." In *Shapeshifting: Transformations in Native American Art*, edited by Karen Kramer Russell, 212–21. New Haven, CT: Yale University Press, 2012.

Cutschall, Colleen. "The Seen and the Unseen to Form the Narrative …" In *Plains Indian Drawings, 1865–1935*, edited by Jane Catherine Berlo, 64–65. New York, NY: Harvey N. Abrams, 1996.

Heap of Birds, Edgar. "Of Circularity and Linearity in the Work of Bear's Heart." In *Plains Indian Drawings, 1865–1935*, edited by Jane Catherine Berlo, 66–67. New York, NY: Harvey N. Abrams, 1996.

Hogan, Linda. *Savings*. Minneapolis, MN: Coffee House Press, 1988.

Horse Capture, Joe D. "Time-Honored Expression: The Knowing of Native Objects," In *Shapeshifting: Transformations in Native American Art*, edited by Karen Kramer Russell, 76–85. New Haven, CT: Yale University Press, 2012.

Horton, Jessica L. "Places to Stand: Art after the American Indian Movement." *Wasafiri: Native North American Literature and Literary Activism* 32, no. 2 (Summer 2017): 23–31.

Lorde, Audre. "Poetry Is Not a Luxury." *Sister Outsider: Essays and Speeches*. Freedom, CA: Crossing Press, 1984. 36–39.

Poitras, Jane Ash. "Paradigms for Hope and Posterity: Woham's Sun Dance Drawing." In *Plains Indian Drawings, 1865–1935*, edited by Jane Catherine Berlo, 68–69. New York, NY: Harvey N. Abrams, 1996.

Contributors

Kimberly Blaeser (Anishinaabe), a Professor of Creative Writing and Native American Literature at University of Wisconsin—Milwaukee, is also a faculty member for Institute of American Indian Arts low rez MFA program in Santa Fe. The author of three poetry collections—most recently *Apprenticed to Justice*—and editor of *Traces in Blood, Bone, and Stone: Contemporary Ojibwe Poetry*, she served as Wisconsin Poet Laureate for 2015–2016. An enrolled member of the Minnesota Chippewa Tribe, Blaeser grew up on the White Earth Reservation.

Gregory A. Cajete (Tewa) is director of Native American studies and a professor in the Division of Language, Literacy and Socio Cultural Studies in the College of Education at the University of New Mexico. Dr. Cajete designs culturally-responsive curricula geared to the special needs and learning styles of Native American students. These curricula are based upon Native American understanding of the "nature of nature" and utilizes this foundation to develop an understanding of the science and artistic thought process as expressed in Indigenous perspectives of the natural world.

Brian S Collier directs the American Indian Catholic Schools Network at the Institute for Educational Initiatives at the University of Notre Dame. He started his teaching career as a volunteer at St. Catherine Indian School in Santa Fe, New Mexico. Collier attended graduate school, earning a Ph.D.

in history, studying why Catholic Indian schools around the United States were closing. Collier had the opportunity to work as a registered lobbyist and learned about the political system and its interactions with educational institutions.

Georges Enderle is the John T. Ryan, Jr. Professor of International Business Ethics at the Mendoza College of Business, University of Notre Dame. His research focuses on the ethics of globalization, wealth-creation, business and human rights, corporate responsibilities of large and small companies, with a view on developments in China. Website: www.nd.edu/genderle.

Four Arrows (Wahinkpe Topa), aka Dr. Don Trent Jacobs (Cherokee), is professor of educational leadership at Fielding Graduate University. Selected as one of 27 "visionaries in education" by AERO Alternative Education Resource Organization (AERO) and recipient of a Martin Springer Institute Moral Courage Award for his activism, he has authored 20 books and numerous chapters and articles on Indigenous worldview, including *Teaching Truly*, selected as one of the century's top twenty progressive education books by the Chicago Wisdom Project.

Eugene Halton is professor of sociology and concurrent professor of American Studies at the University of Notre Dame. His most recent book is *From the Axial Age to the Moral Revolution*. His earlier books include *The Great Brain Suck, Bereft of Reason, Meaning and Modernity*, and *The Meaning of Things*.

Bruce E. Johansen is Frederick W. Kayser Professor of Communication and Native American Studies at the University of Nebraska at Omaha, where he has been teaching and writing since 1982. He has authored 47 published books, most recently a three-volume set, *Climate Change: An Encyclopedia of Science, Society, and Solutions* (2017). Johansen holds the University of Nebraska award for Outstanding Research and Creative Activity (ORCA), the state system's highest faculty recognition.

Steve J. Langdon, professor emeritus of anthropology at the University of Alaska Anchorage, specializes in understanding Tlingit and Haida Indian culture in southeast Alaska, focusing on the complex patterns of engagement with salmon channeled by the Salmon Boy mythic charter/covenant taught to all Tlingit as children. He is currently working on publication of a book length treatment of Tlingit salmon-relations and has developed exhibits demonstrating Tlingit traditional salmon fishing technologies that will be housed in the Soboleff Center of the Sealaska Heritage Institute located in Juneau, Alaska.

Barbara Alice Mann (Seneca), is professor of humanities in the Jesup Scott Honors College at the University of Toledo, in Toledo, Ohio, USA. The most recent of her thirteen published books are *Spirits of Blood, Spirits of Breath* (2016) and

The Tainted Gift (2009), while her "classic" *Iroquoian Women* (2000) remains in print. With hundreds of published chapters and articles, she is currently working on two more books, including "President by Massacre," to be out in 2019.

Darcia Narvaez (Puerto Rican) is professor of psychology at the University of Notre Dame who specializes in moral development and human flourishing. Her book, *Neurobiology and the Development of Human Morality: Evolution, Culture and Wisdom* won the 2015 William James Book Award from the American Psychological Association and the 2017 Expanded Reason Award. In it she explains how the Indigenous worldview and communal morality are shaped by early experience and can be restored by revamping one's neurobiology through practices of self-calming, social joy and ecological attachment.

Penny Spikins is a senior lecturer on the archaeology of human origins in the Department of Archaeology, University of York, UK. She is particularly interested in how archaeological evidence can contribute to our understanding of human cognitive and social evolution. She has published papers on topics such as the evolution of compassion, egalitarian dynamics, childcare and emotional commitments as well as a recent volume *How Compassion Made Us Human* (Pen and Sword).

Sandra Waddock is Galligan Chair of Strategy, Carroll School Scholar of Corporate Responsibility, and Professor of Management at Boston College's Carroll School of Management. Author of more than 150 papers and 13 books, she has received numerous awards, including the 2016 Lifetime Achievement Award for contributions to corporate social responsibility from Humboldt University's International CSR conference. Current research interests include large system change, memes and narratives, and intellectual shamanism. Her latest book is *Healing the World* (2017, Greenleaf/Routledge).

White Standing Buffalo (aka Tom McCallum) (Cree/Metis) is honored to serve the people as a Sundance Chief and a holder of other ceremonial bundles passed on to him by respected Elders and Ceremonialists. In Canada, he is contracted to provide spiritual guidance and ceremony to federal inmates. As an Elder and storyteller, he has worked in the healing field for decades, providing cultural counseling and traditional healing ceremonies for those seeking guidance on exploring, or reconnecting with, the Aboriginal worldview.

Jon Young is a founder of the 8 Shields Institute, the Nature Connection Mentoring Foundation, the Bird Language Leaders, and the Wilderness Awareness School, he has pioneered nature connection models for over 30 years. He is the author of *What the Robin Knows: How Birds Reveal the Secrets of the Natural World*, *The Kamana Naturalist Training Program*, *Animal Tracking Basics* and *Coyote's Guide to Connecting with Nature*.

Index